T0259724

Acoustic Analysis and Design of Short Elliptical
End-Chamber Mufflers

Akhilesh Mimani

Acoustic Analysis and Design of Short Elliptical End-Chamber Mufflers

Akhilesh Mimani
Department of Mechanical Engineering
Indian Institute of Technology Kanpur
Kanpur, Uttar Pradesh, India

ISBN 978-981-10-4827-2   ISBN 978-981-10-4828-9   (eBook)
https://doi.org/10.1007/978-981-10-4828-9

This Springer imprint is published by the registered company Springer Nature Singapore Pte Ltd.
The registered company address is: 152 Beach Road, #21-01/04 Gateway East, Singapore 189721, Singapore

Foreword

The automobile engine exhaust mufflers that are located under the vehicle are often elliptical in shape as a consequence of lowering the center of gravity of the vehicle in order to ensure mechanical stability. The desirability of minimum volume, weight and cost of the muffler necessitates acoustically small end-chambers in the flow-reversal type muffler configurations: Plane wave propagation breaks down and higher-order modes start propagating in these small end-chambers right from frequencies of the order of the firing frequency of the high-speed multi-cylinder automotive engines. This calls for 3-D analysis of the elliptical exhaust mufflers. The numerical finite element method is too cumbersome and costly to be of any use to muffler designers. Dr. Akhilesh Mimani steps in here to offer this Springer Monograph on acoustic analysis and design of short elliptical end-chamber mufflers.

Based on his doctoral work at the Facility for Research in Technical Acoustics, Indian Institute of Science, Bangalore, and post-doctoral work at leading Australian Universities during the last 14 years, this monograph makes use of 3-D Green's function derived in terms of Mathieu functions, modal summation, impedance matrix and a uniform piston-driven model. Probably for the first time, the parametric zeros and non-dimensional resonance frequencies of the higher-order transverse modes of a rigid-wall elliptical waveguide for a complete range of aspect-ratio have been calculated and tabulated in this monograph.

An ingenious feature of Mimani's monograph is formulation of a set of comprehensive guidelines that suggest optimal inlet/outlet port locations which would suppress the propagation of certain higher-order transverse modes, thereby yielding a broadband attenuation despite a small expansion volume. This feature is very useful for designing short elliptical (and circular) mufflers of the flow-reversal type that are often used as end-chambers in modern-day automotive exhaust systems.

Finally, a useful feature of this monograph is that each chapter is written so as to be independently readable, with a comprehensive list of up-to-date references.

In conclusion, this monograph has great archival value and is recommended strongly to all acousticians as well as muffler designers.

Dr. M. L. Munjal
Professor (Emeritus)
Indian Institute of Science
Bengaluru, India

Preface

This monograph presents a three-dimensional analysis of acoustic wave propagation in an elliptical waveguide and applies the equations and concepts to design axially short elliptical end-chamber muffler configurations which are an important component of a complex multi-pass muffler used in a modern-day automotive exhaust system. A general solution of the Helmholtz equation in elliptical cylindrical coordinates is presented in terms of the Mathieu and modified Mathieu modal functions. This is followed by the tabulation and analysis, for the first time, of the non-dimensional resonance frequencies of the transverse modes of a rigid-wall elliptical waveguide for a complete range of aspect-ratio. The modal shape patterns of the first few circumferential, radial and cross-modes are examined with particular attention to the pressure nodal ellipses and hyperbolae. An analytical formulation is then outlined for characterizing a single-inlet and single-outlet elliptical muffler with the inlet located on the end face and the outlet located either on the end face or side surface. The ensuing chapter is devoted toward analyzing the transmission loss (TL) performance of different short end-chamber mufflers, namely (a) the straight-flow configuration having ports located on the opposite face, (b) the flow-reversal configuration with ports located on the same end face and (c) configuration with inlet port on the end face and outlet on the side surface. Design guidelines are formulated in terms of the optimal location of inlet and outlet ports which suppresses the deteriorating influence of certain higher-order modes, thereby delivering a broadband TL performance. Directions for future work are discussed toward the end.

In summary, this monograph is a one-stop solution for a practicing automotive engineer designing mufflers, for an applied mathematician studying wave propagation in elliptical geometries, and also as a niche area within noise control engineering.

Kanpur, India   Akhilesh Mimani

Acknowledgements

I would like to acknowledge all academics and mentors located across different universities in India as well as Australia with whom I have had the privilege to work, or have interactions with. In particular, I would like to thank Prof. M. L. Munjal, at Indian Institute of Science (IISc), Bengaluru who as a doctoral thesis supervisor not only introduced me to muffler acoustics several years ago, but in due course also encouraged me to write this book. The acknowledgment is rather incomplete without the special mention of Prof. Anindya Chatterjee, a senior colleague at IIT Kanpur from whom I first learned about the Mathieu functions when I was pursuing my doctoral degree at IISc. I would also like to thank Prof. Ray Kirby at the University of Technology, Sydney with whom I had an opportunity to work and benefit through technical discussions. Lastly, but certainly not the least, I would like to express my gratitude to Prof. Nicole Kessissoglou of UNSW Sydney who was helpful in many ways during my stay in Sydney.

This monograph is dedicated to my wife Shruti—I am grateful for her constant support over all these years.

Kanpur, Uttar Pradesh, India   Akhilesh Mimani
October 2020   amimani@iitk.ac.in

Contents

About the Author

Akhilesh Mimani is an Assistant Professor at Department of Mechanical Engineering, Indian Institute of Technology Kanpur (IITK), Uttar Pradesh India. Akhilesh received his Ph.D. (2012) in Mechanical Engineering from the prestigious Indian Institute of Science, Bangalore specializing in muffler and duct acoustics. He has completed research associate positions at The University of Adelaide (2016), University of New South Wales Sydney (2017) and University of Technology Sydney (2018) before joining IITK in 2018. Akhilesh has 10 years of experience in the analysis and design of mufflers for automotive and industrial applications which includes consultancy. His other research interests include computational and experimental aeroacoustics, array processing techniques such as time-reversal and beamforming for source localization, and application of the finite element method to acoustic wave propagation problems. Akhilesh's work has been published in reputed international journals and conference proceedings, and he has received funding from both government and industry, in addition to travel grants.

Chapter 1
Introduction

1.1 Elliptical Mufflers for Automotive Exhaust Application: Motivation

Exhaust noise of reciprocating internal combustion (IC) engines is certainly one of the biggest pollutants of the present-day urban environment. Fortunately, the use of a well-designed muffler (also known as a silencer) can significantly mitigate this problem by reducing the noise from IC engines [1]. Indeed, all automotive engines are invariably provided with exhaust mufflers, the theory and design practice of which is now over a hundred years! A primary design requirement of an automotive exhaust muffler is obtaining an adequate insertion loss so that the exhaust noise is reduced to the level of the noise from other components of the engine, or as required by the environmental noise pollution limits. Generally speaking, the amount of acoustic attenuation produced by a muffler is proportional to the expansion chamber volume in the low-frequency range which is particularly important because most of the engine noise is limited to the firing frequency and the first few harmonics. The necessity to have a sufficiently large volume is reinforced by the need to have a large expansion ratio, i.e., a sharp impedance mismatch. Table 5.1 of Ref. [2] (reproduced from Bies and Hansen [3]) gives a good estimate of how important the muffler volume is in context with the attenuation produced at different octave band frequency bands; as a rule of thumb, small and large mufflers are approximately characterized by 5 and 15 times the piston displacement capacity of the engine, respectively. Unfortunately, however, the clearing space beneath the automobile body, where the muffler is typically located, is kept small because the stability of a vehicle requires a low center of gravity. Therefore, the space constraint along the vertically downward direction coupled with an additional essential requirement that the muffler shell should not touch the ground (especially on a rough terrain) often leads to the use of silencing chambers having a non-circular shape. Additionally, manufacturing defects or constraints might also force one to use non-axisymmetric chambers. In view of these design constraints, an elliptical and for that matter, a flat-oval chamber is popularly used in modern automobile exhaust systems, see, e.g., Figure 1.1a. Elliptical

A. Mimani, *Acoustic Analysis and Design of Short Elliptical End-Chamber Mufflers*,
https://doi.org/10.1007/978-981-10-4828-9_1

1

Fig. 1.1 a A photograph illustrating the use of elliptical mufflers in modern-day automotive exhaust system. [The source of this photograph is https:// en.wikipedia.org/wiki/Muf fler#/media/File:Exhaust_p ipe_muffler.JPG. This file is licensed under the Creative Commons Attribution 2.5 Generic license and it is free to be shared (copied, distributed and transmitted)]. **b** Interior of a typical three-pass perforated tube elliptical muffler showing different elements such as perforated tubes, annular cavity filled with a sound absorbent (dissipative) material, short end-chambers at the front and rear, and baffles

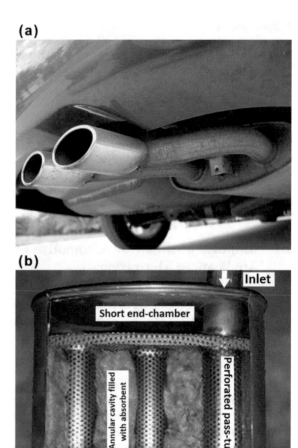

chamber with a straight-through flow configuration also finds use in the exhaust system of two-wheelers.

Among other requirements, the 'breakout' noise from muffler shell must also be minimized, i.e., the shell must have a high transverse transmission loss (TTL) so that net transmission loss is nearly equal to the axial transmission loss (TL). While a perfectly circular muffler shell has a high TTL owing to its rigidity, departures from axisymmetry adversely impact the TTL or increase breakout noise emission as may be seen from the results for square and rectangular ducts used in heating, ventilation and air-conditioning (HVAC) systems [1, 4]. Munjal and co-workers [5], however, report that the problem is not as severe for an elliptical muffler shell whose TTL performance throughout the frequency range is somewhat intermediate between the circular chamber and the flat-plate or flat-oval geometry which is by far, the worse. This is yet another appealing reason for use of elliptical chambers in automotive exhaust system.

1.2   State of the Art and Scope

The paper by Lowson and Baskaran [6] was probably one of the first published works dealing with acoustic wave propagation inside a rigid-wall elliptical cylindrical waveguide in terms of the Mathieu modal functions and proved to be the beginning of analytical investigations for elliptical acoustic cavities [7, 8], lined ducts [9] and mufflers [10–17].

Denia et al. [10] presented a point-source or Green's function method based on the Mathieu modal summation to analyze the acoustic attenuation performance of elliptical chamber mufflers of the simple expansion and flow-reversal type configurations. Their work was the first one to employ an analytical method for carrying out a parametric investigation on the effect of the chamber length and relative angular locations of inlet and outlet as well as eccentricity on the TL performance. Comparisons were provided with finite element (FE) results and experimental measurements which showed a good agreement. In addition, they also developed polynomial expressions to evaluate the cut-off frequencies of the first few higher-order modes and the optimum location of the outlet pipe in order to increase the effective attenuation range. In a following work, Denia et al. [11] investigated the TL performance of a single-inlet and double-outlet elliptical chambers by applying the more accurate analytical mode-matching (AMM) method at the interface of the elliptical chamber and circular ports. The AMM method allowed them to conduct parametric studies which revealed measures to improve the attenuation performance by optimal port location. Mimani and Munjal [12] used the uniform piston-driven model and the Mathieu function expansions to characterize an elliptical muffler having an end-inlet and a side-outlet in terms of the impedance $[\mathbf{Z}]$ matrix parameters whereby an excellent agreement was observed between their semi-analytical and FE predictions. The authors further extended their analysis to single-inlet and double-outlet elliptical mufflers with arbitrary location of inlet and outlet ports, either on the end face or on the side surface [13]. Parametric studies revealed (axially) long and short chamber configurations with certain locations of the inlet and outlet ports that deliver a broadband attenuation, thereby allowing them to come up with some design guidelines. The semi-analytical piston-driven model approach was later used to evaluate the end-correction in long flow-reversal end-chamber configurations for which the ports are flush-mounted on an end face [14]. Sohei et al. [15] derived the four-pole parameters of an elliptical muffler chamber having a central perforated tube which was modeled as piston-driven rigid tube. The 3-D acoustic pressure field inside the muffler chamber was expressed as modal summation involving even and odd Mathieu functions, and appropriate boundary conditions were considered at the inlet and outlet pistons as well as the perforate-chamber interface.

Banerjee and Jacobi [16] employed a regular perturbation method to determine the TL performance of moderately elliptical mufflers of simple expansion and flow-reversal type and studied the effect of various inlet/outlet orientations. Their results were found to agree well with the Green's function analytical method of Denia et al. [11] as well as FE prediction. In a follow-up paper, they analyzed a low eccentricity elliptical muffler with concentric extended-inlet and extended-outlet system using the perturbation analysis; they acknowledged that the annular regions formed by the inlet and outlet extensions are non-confocal making it impossible to implement the boundary conditions exactly; thus, the analytical modes could not be obtained. The perturbation method was, therefore, used to obtain resonance frequencies and mode shapes in the annular regions followed by a mode-matching procedure at the inlet and outlet extensions. Some design measures were suggested to enhance the attenuation graphs by tuning the extension lengths.

With the availability of inexpensive computing power, however, a much more attractive option for the analysis of elliptical automotive mufflers is certainly, the FE-based numerical mode-matching (NMM) technique. For example, Kirby [18, 19] used the FE-based NMM to evaluate the TL performance of a straight-through flow elliptical muffler having a fully perforated concentric airway where the annular region was filled with a sound absorptive or dissipative material. Such configurations produce a minimum back-pressure, and owing to the presence of sound absorbent, it delivers a good attenuation performance in the high-frequency range. The NMM has also been used to evaluate the TL performance of elliptical and flat-oval mufflers having an extended-inlet and extended-outlet where the extended tubes need not be concentric [20, 21]. Additionally, large industrial silencers of elliptical cross-sections with a partially perforated concentric or eccentric airway have also been analyzed using NMM, and some guidelines were suggested to improve the TL performance by carefully selecting the extension lengths [22].

Elliptical cross-sections are also quite common in a three-pass perforated (TPP) muffler configuration which are popularly used in modern-day automotive silencing system [23, 24]. Figure 1.1b shows the interior of a typical TPP muffler where the important constituting elements are shown, namely the fully perforated tubes, annular cavity of the middle chamber filled with a sound-absorbing material and the flow-reversal end-chambers which are so named because they play the role of reversing the mean flow direction forcing it to pass through multiple perforated pipes. This allows for a greater interaction of the waves with the dissipative annular cavity leading to a higher attenuation within the constraints of limited space. From an acoustic point-of-view, the end-chambers are essentially expansion chambers which deliver attenuation either by reflecting back a significant portion of incident waves or they act like acoustic resonators which produce an attenuation peak at certain (resonance) frequencies. Unfortunately, however, due to lack of sufficient space, the length of end-chambers is usually small as shown in Fig. 1.1(b) which means less expansion volume or equivalently, less attenuation in the low-frequency range. The acoustical behavior of such short end-chambers is known to be drastically different from that of a long chamber because the higher-order evanescent modes generated at the ports do not decay sufficiently and, in fact, completely dominate the axial modes [25–29]. Young and Crocker [25] used FEA to study the acoustic behavior of short flow-reversal elliptical chambers and showed that such chambers have two completely different TL characteristics: One is similar to that of the simple expansion chamber and the other comparable to that of a side-branch resonator. In either cases, it suggests the acoustic modes are predominantly oriented along the major-axis. Mimani and Munjal [26, 27] confirmed this hypothesis; they modeled short elliptical (and circular) chambers using the 1-D transverse plane wave theory which assumes planar wave propagation along the major-axis. This approach yielded satisfactory results up to the resonance frequencies of the (2,1) even and (1,0) circumferential modes for elliptical and circular chambers, respectively. Selamet and Ji [28] presented a 3-D analytical approach based on mode-matching to determine the TL performance of circular flow-reversal muffler configuration; for short chambers, it was shown that when one of the end ports is centered while the other is offset on the pressure nodal circle of the (0,1) radial mode, a broadband attenuation performance is obtained up to the onset of the (0,2) radial mode. Similar design considerations are also presented in Refs. [12, 29] for short elliptical chambers with an end-inlet and side-outlet.

The primary objective of this monograph is then to present a comprehensive account of the acoustic analysis and design of short end-chamber mufflers having an elliptical or circular cross-section. We consider a 3-D semi-analytical approach based on the modal summation and uniform piston-driven model [12–14] to analyze and eventually recommended guidelines in terms of possible optimal locations/arrangements of ports for such muffler configurations. The analytical consideration of the rigid-wall acoustic modes in the elliptical chamber, however, requires the numerical computation of parametric zeros of the derivative of the modified or radial Mathieu functions evaluated at the elliptical boundary. The underlying idea is to first determine the resonance frequencies and mode shapes of the transverse modes for a given aspect-ratio. While this appears straightforward, unfortunately, for a muffler analyst, readily available tables which systematically document the resonance frequencies of first several modes of an elliptical chamber of arbitrary aspect-ratio are rather difficult to get hold of. The difficulty is most probably because *'the making of numerical tables usually receives little encouragement, while the thanks offered by the user are parsimonious rather than plentiful'* to quote McLachlan [30]. To resolve this issue, we present tables of resonance frequencies (and parametric zeros) corresponding to the Neumann condition for aspect-ratio ranging from highly eccentric sections, i.e., flattened ellipses to a perfect circular section—documentation of these tables also falls within the scope of this monograph.

1.2.1 Book Outline

The monograph is organized as follows. Chapter 2 presents the analytical solution of the three-dimensional (3-D) acoustic field inside an infinite elliptical cylindrical waveguide carrying a uniform mean flow expressed in terms of the Mathieu and modified Mathieu functions. The non-dimensional resonance frequency corresponding to the rigid-wall modes are documented for aspect-ratio ranging from a highly eccentric elliptical to a perfect circular section. This is followed by interpolation formulae, mode shape analysis and their possible implication on elliptical muffler design. Chapter 3 presents the required theoretical formulation based on the modal solution to characterize an elliptical muffler with end-inlet and end-/side-outlet ports. Chapter 4 builds upon the foundation provided by the preceding chapters; it presents a detailed TL analysis of the short elliptical and circular end-chamber mufflers with different arrangements of inlet and outlet ports. An important outcome for an engineer designing automotive mufflers includes the guidelines on optimally locating the inlet/outlet ports of a short elliptical and circular end-chamber muffler which deliver a broadband TL performance over a maximum frequency range. Chapter 5 summarizes the contributions and presents some ideas for extending the present work to suit the analysis of more complicated muffler designs.

References

1. M.L. Munjal, *Acoustics of Ducts and Mufflers*, 2nd edn. (Wiley, Chichester, UK, 2014)
2. M.L. Munjal, *Noise and Vibration Control*. IISc Lecture Note Series (World Scientific, Singapore, 2013)
3. D.A. Bies, C.H. Hansen, *Engineering Noise Control*, 3rd edn. (Spon Press, New York, 2003)
4. B. Venkatesham, M. Tiwari, M.L. Munjal, Analytical prediction of break-out noise from a reactive rectangular plenum with four flexible walls. J. Acoust. Soc. Am. **128**, 1789–1799 (2010)
5. M.L. Munjal, G.S.H. Gowtham, B. Venkatesham, H.K.M. Reddy, Prediction of breakout noise from an elliptical duct of finite length. Noise Control. Eng. J. **58**, 319–327 (2010)
6. M.V. Lowson, S. Baskaran, Propagation of sound in elliptic ducts. J. Sound Vib. **38**, 185–194 (1975)
7. K. Hong, J. Kim, Natural mode analysis of hollow and annular elliptical cylindrical cavities. J. Sound Vib. **183**, 327–351 (1995)
8. W.M. Lee, Natural mode analysis of an acoustic cavity with multiple elliptical boundaries by using the collocation multipole method, J. Sound Vib. **330** 4915–4929 (2011)
9. J.M.G.S. Oliveira, P.J.S. Gil, Sound propagation in acoustically lined elliptical ducts, J. Sound Vib. **333** 3743–3758 (2014)
10. F.D. Denia, J. Albelda, F.J. Fuenmayor, A.J. Torregrosa, Acoustic behaviour of elliptical chamber mufflers, J. Sound Vib. **241**, 401–421(2001)
11. F.D. Denia, L. Baeza, J. Albelda, F.J. Fuenmayor, in Acoustic behaviour of elliptical mufflers with single-inlet and double-outlet. Tenth International Congress on Sound and Vibration (7–10 July Stockholm, Sweden, 2003)
12. A. Mimani, M.L. Munjal, 3-D acoustic analysis of elliptical chamber mufflers having an end inlet and a side outlet: an impedance matrix approach. Wave Motion **49**, 271–295 (2012)

13. A. Mimani, M.L. Munjal, Acoustical behavior of single inlet and multiple outlet elliptical cylindrical chamber muffler. Noise Control Eng. J. **60**, 605–626 (2012)
14. A. Mimani, M.L. Munjal, Acoustic end-correction in a flow-reversal end chamber muffler: A semi-analytical approach. J. Comput. Acoust. **24**, 1650004 (2016)
15. N. Sohei, N. Tsuyoshi, Y. Takashi, Acoustic analysis of elliptical muffler chamber having a perforated pipe. J. Sound Vib. **297**, 761–773 (2006)
16. S. Banerjee, A.M. Jacobi, Analysis of sound attenuation in elliptical chamber mufflers by using Green's function. ASME Paper No. IMECE2011-65345
17. S. Banerjee, A.M. Jacobi, Analytical prediction of transmission loss in distorted circular chamber mufflers with extended inlet/outlet ports by using a regular perturbation method. J. Vib. Acoust. **37**, 061002 (2015)
18. R. Kirby, Transmission loss predictions for dissipative silencers of arbitrary cross section in the presence of mean flow. J. Acoust. Soc. Am. **114**, 200–209 (2003)
19. R. Kirby, A comparison between analytic and numerical methods for modelling automotive dissipative silencers with mean flow. J. Sound Vib. **325**, 565–582 (2009)
20. Z. Fang, Z.L. Ji, Acoustic attenuation analysis of expansion chambers with extended inlet/outlet. Noise Control Eng. J. **61**, 240–249 (2013)
21. Z. Fang, Z.L. Ji, Finite element analysis of transversal modes and acoustic attenuation characteristics of perforated tube silencer. Noise Control Eng. J. **60**, 340–349 (2012)
22. A. Mimani, R. Kirby, in Design of large reactive silencers for industrial applications. *Proc. of InterNoise* (26–29 Aug 2018, Chicago, USA)
23. H. Huang, Z. Ji, Z. Li, Influence of perforation and sound-absorbing material filling on acoustic attenuation performance of three-pass perforated mufflers. Adv. Mech. Eng. **10**, 1–11 (2018)
24. M.L. Munjal, Analysis of a flush-tube three-pass perforated element muffler by means of transfer matrices. Int. J. Acoust. Vib. **2**, 63–68 (1999)
25. C.I.J. Young, M.J. Crocker, Acoustical analysis, testing and design of flow-reversing muffler chambers. J. Acoust. Soc. Am. **60**, 1111–1118 (1976)
26. A. Mimani, M.L. Munjal, Transverse plane wave analysis of short elliptical chamber mufflers: An analytical approach. J. Sound Vib. **330**, 1472–1489 (2011)
27. A. Mimani, M.L. Munjal, Transverse plane-wave analysis of short elliptical end-chamber and expansion-chamber mufflers. Int. J. Acoust. Vib. **15**, 24–38 (2010)
28. A. Selamet, Z.L. Ji, Acoustic attenuation performance of circular flow-reversing chambers. J. Acoust. Soc. Am. **104**, 2867–2877 (1998)
29. A. Selamet, F.D. Denia, Acoustic behavior of short elliptical chambers with end central inlet and end offset or side outlet. J. Sound Vib. **245**, 953–959 (2001)
30. N.W. McLachlan, *Theory and Application of Mathieu Functions*, Chap. 1, pg. 7 (Oxford University Press, London, 1947)

Chapter 2
Acoustic Wave Propagation in an Elliptical Cylindrical Waveguide

2.1 Introduction: A Brief Review on Mathieu Functions and Objectives

Mathieu functions are named after E. Mathieu who in 1868 first introduced them when he determined the natural vibrational modes of a stretched elliptical membrane in his seminal work '*Memoire sur le movement vibratoire d'une membrane de forme elliptique*' [1]. They are the eigenfunction solutions, when the wave equation is separated in the elliptical co-ordinates ξ and η, and are the counterparts of the more commonly encountered ordinary Bessel and trigonometric functions that arise while solving physical problems in radial co-ordinates. Following the appearance of Mathieu's work, a decade elapsed before Heine [2] published his work in which he defined the periodic Mathieu functions of integer order as Fourier sine and cosine series, without evaluating the corresponding coefficients; obtained a transcendental equation for the characteristic numbers; and demonstrated that one set of periodic functions of integer order could be expanded in a series of Bessel functions. In 1883, Floquet published a general treatment of linear differential equations with periodic coefficients [3], of which the Mathieu's and Hill's equation [4] were special cases. However, the work by Floquet and Hill both played an important role in the development of Mathieu functions. The interested reader is referred to the book by N. W. McLachlan [5] in which a very extensive mathematical treatment is provided for numerically computing both angular and radial Mathieu functions, including a range of asymptotic formulae. The first chapter of this classical text provides a detailed historical introduction and development of Mathieu functions, along with an updated bibliography through 1947. In the second half of this book, the application of Mathieu functions to different physical problems is presented which include vibrational systems (membranes and plates), electrical and thermal diffusion, electromagnetic waveguides , and diffraction of sound and electromagnetic waves. The other books include Meixner and Schäfke [6] and the chapter by Blanch in the handbook by Abramowitz and Stegun [7].

The very early investigations focused on finding the characteristic values a in the form of infinite continued fractions and introduction of the $a - q$ chart, and

© The Author(s), under exclusive license to Springer Nature Singapore Pte Ltd. 2021
A. Mimani, *Acoustic Analysis and Design of Short Elliptical End-Chamber Mufflers*,
https://doi.org/10.1007/978-981-10-4828-9_2

tabulation of the first few zeros of the angular Mathieu functions of integral order, see Ince [8]. (Here, q is the parameter associated with the periodic function in the Mathieu functions; when it is zero, one obtains the familiar trigonometric functions as solutions.) Published literature also exists which presents the tabulated values of the modified Mathieu functions for the first few orders and a range of arguments [9], given that their computation was not a straightforward task at least about 50 years ago. In fact, given the computational constraints back in the late 1920s, the development of asymptotic formulae for Mathieu functions was in vogue [10] which, in fact, continues to attract the attention of the present-day mathematicians and engineers alike, see the paper by Canosa [11], and the relatively recent ones by Frenkel and Portugal [12] as well as Alhargan [13] who provided accurate asymptotic formulae for the computation of the Mathieu characteristic numbers by the introduction of a new normalization scheme. Indeed, a library of C++ routines was also developed by Alhargan [14, 15] which is available from the ACM Digital Library to evaluate both angular and radial Mathieu functions for orders up to 200 and q-values up to 160,000.

It was only with the availability of computing power that works on numerical computation of Mathieu functions gained momentum from the late 1960s. Clemm [16] published several pioneering Fortran routines to evaluate the characteristic values by obtaining a rough approximations based on the results from curve-fitting of the available tabulation followed by iteration using a Newton's method. While the method produced accurate values of the radial and angular Mathieu functions, the main criticism was the dependence on previously tabulated values. Rengarajan et al. [17] refined the Clemm's method by computing the Bessel functions in a separate subroutine and by taking advantage of the normalization scheme defined by McLachlan [5]. Their routines had an accuracy of seven decimal places for the first 11 orders of the angular and radial Mathieu functions as well as their derivatives but only for small values of q. Toyama et al. [18] show that by method of truncating the continued fractions, the characteristic values were obtained for a much large range of q given by 0–30. Leeb [19] showed that the characteristic values can be computed by solving the eigenvalue problem of asymmetrical tridiagonal matrix system up to a much larger range of q given by 0–250. Shirts [20] reported two Fortran routines which calculate the characteristic numbers and Mathieu functions for fractional as well as integer order, and a very high accuracy of the order 10^{-14} was obtained. The first routine was based on the standard eigenvalue problem of the tridiagonal matrix, whereas, the second employed the usual continued fraction technique to find the characteristic numbers. Stamnes et al. [21] also presented a method based on formulating the calculation as an eigenvalue problem to compute the Mathieu functions. Zhang and Jin [22] present useful Fortran-based routines for computation of these functions. In the doctoral work by Gutiérrez-Vega [23], the angular Mathieu functions were expanded in series of even and odd trigonometric functions, and the recurrence relationship was obtained between the expansion coefficients of each series, which were found out using the continued fractions approach as well as by solving the eigenvalue problem of the symmetric tridiagonal truncated matrix. The radial Mathieu functions were computed using the rapidly converging series of product of Bessel functions

[5]. The MATLAB toolbox (available for free download) for computing the Mathieu functions written by Cojocaru [24] follows the same procedure and symbolic notation as those described in Ref. [23]—the characteristic or eigen values a and the normalized expansion coefficients or the eigenvectors, angular and radial Mathieu functions were the final outputs of the functions constituting the toolbox.

While the techniques for computation of Mathieu functions of arbitrary large order and arguments have drawn the attention of the scientific community, of equal importance are the papers that are dedicated to the computation of resonance frequencies of engineering systems with an elliptical boundary such as a clamped membrane or a rigid-wall waveguide. B. A. Troesch and H. R. Troesch [25] used Newton's method to find the roots from the Fourier series expressions for the appropriate modified Mathieu functions and list asymptotic approximations for the larger roots. More precisely, they computed the first few resonance frequencies and corresponding mode shapes of a clamped vibrating elliptical membrane. The variation of the non-dimensional resonance frequencies of the even and odd modes was graphically shown as a function of eccentricity of constant area ellipses wherein it was found that parameters are remarkably stationary over a range of modes and eccentricity. They did not, however, provide the data as tabulated values which would have been more readily accessible for a quick reference. The paper by Chen et al. [26] relates the modern numerical evaluation to some of the asymptotic predictions by Keller and Rubinow [27] and also draws attention to the three distinct classes of high-frequency modes that can exist on a clamped elliptic membrane, namely the whispering gallery, bouncing ball and focus modes. The tutorial by Gutiérrez-Vega et al. [28] summarizes several applications of Mathieu's differential equation through visualization of the different modes types mentioned above. Ancey et al. [29] were able to compute accurate values for the resonance frequency of whispering gallery class of modes pertaining to the Dirichlet condition, by direct application of the Langer–Olver asymptotic method to the Mathieu's differential equation. In addition to the algorithms based on solving the eigenvalue problem to compute the resonance frequencies via a direct evaluation of the Mathieu functions, some approximate treatments have also been presented in the literature to compute the resonance frequencies (and mode shapes) [30, 31]. Akulenko and Nesterov [30] used a Rayleigh–Ritz variational formulation to compute the lower resonant frequencies for ellipses ranging from nearly circular to high eccentric. A noteworthy contribution along these lines is due to Wilson and Scharstein [31] who proposed a Galerkin formulation to directly compute the resonance frequencies of a membrane without invoking Mathieu functions. They presented the resonance frequencies of the first 100 even and odd modes corresponding to the soft-wall (Dircihlet) and hard-wall (or Neumann) boundary conditions for an ellipse with aspect-ratio equal to 0.5. The tabulated values were shown to be in an excellent agreement with those computed by directly evaluating the Mathieu functions or the use of 2-D finite element (FE) discretization to the elliptical domain. They also presented the whispering gallery, focus and bouncing ball mode shapes, which in this work are referred to as odd circumferential, even circumferential and radial modes, respectively, as this terminology is common in the field of duct and muffler acoustics which is the focus of this work.

In parallel with the work on vibrational modes of an elliptical membranes, wave propagation in elliptical cylindrical ducts (waveguides) which is often associated with the rigid-wall boundary condition has also received significant attention [32–35]. The earliest papers are by Chu [32] and Daymond [33] who focused their investigations to the study of the lowest-order resonance frequency which means that up to their work, very little was known about the behavior of higher-order elliptical duct modes. The complete acoustic pressure (or electromagnetic wave) field in a rigid-wall elliptical waveguide is, however, expressed as a summation of higher-order modes including the fundamental or plane wave mode [35–38]. To obtain such a solution, we must first find the transverse non-dimensional resonance frequencies of acoustic modes in a rigid-wall elliptical waveguide leading to the problem of computing the roots or parametric zeros of the derivative of radial Mathieu functions evaluated at the elliptical boundary. The paper by Lowson and Baskaran [35] was probably the first one to analyze the propagation of higher-order acoustic modes in a rigid-wall elliptical duct in terms of the Mathieu functions. Indeed, their work, perhaps, was also the first attempt for systematic tabulation of the 'exact' non-dimensional cut-off wavenumbers (resonance frequencies) of the first 15 even and odd circumferential modes for a range of eccentricity values $e = 0, 0.1, 0.2(0.1), 0.9, 0.95$. It is important to note that their tabulated values pertained to ellipses of equal area. As a result, regardless of eccentricity, the cut-on frequencies of the higher-order even and odd modes were nearly the same and highly comparable with that of a circular duct of the same area. This implies that if area of a circular duct and a highly eccentric elliptical duct are same, the cut-on frequencies of their counterpart modes differ only marginally. However, in order to better study the effect of eccentricity (aspect-ratio) on variation of cut-on frequencies, and in particular, to bring out the difference in excitation behavior of the even and odd modes, it may be desirable to rescale these values by keeping the semi-major-axis constant. Furthermore, they did not document the cut-on frequencies of purely radial modes and the cross-modes. Hong and Kim [39] obtained analytical solutions for natural frequencies and mode shapes of elliptical cylindrical acoustic cavities by solving the wave equation in elliptical cylindrical coordinates. They considered both hollow and confocal annular elliptical cylindrical cavities, and the first few resonance frequencies and mode shapes pertained to cavities of the size typically used as a hermetic compressor shell or a small automotive muffler. The results for chambers of a given eccentricity were compared with those of a circular cylindrical cavity having equal cross-sectional area and volume. Lee [40] solved the acoustic eigenvalue problem of elliptical cylindrical cavities having multiple elliptical cylinders based on a multi-pole expansion in terms of the angular and radial Mathieu functions and applying the concepts of directional derivative, collocation method and singular value decomposition.

For computing the impedance [**Z**] matrix parameters of reactive (non-dissipative) elliptical chamber mufflers analytically in terms of the modal summation and evaluating the attenuation performance, it is necessary to determine in advance, the non-dimensional resonance frequencies of all transverse mode types, namely the purely radial modes, the even and odd circumferential modes as well as the even and odd cross-modes. Wilson and Scharstein [31], Denia et al. [36] as well as the present author [37, 38] published such a data-set of transverse resonance frequencies for the first few orders of Mathieu functions; however, these values pertained to only specific aspect-ratio. Generally speaking, it is difficult to find in the literature, a systematic documentation of the non-dimensional transverse resonance frequencies of a rigid-wall elliptical duct for a complete range of aspect-ratio or eccentricity, although this data is crucial for computing the attenuation performance of an elliptical muffler as well as for studying the acoustical properties of elliptical enclosures found in several engineering applications.

In light of the background provided above, we take upon ourselves the task of tabulating the parametric zeros and the corresponding non-dimensional resonance frequencies of transverse modes of a rigid-wall elliptical duct with aspect-ratio ranging from $D_2/D_1 = 0.01$ (highly eccentric or flattened) to $D_2/D_1 = 1$ (Circle), for which these values are easily accessible [41, 42]. The chapter begins with the 3-D Helmholtz equation in elliptical cylindrical co-ordinates which governs acoustic wave propagation in an elliptical duct carrying a uniform mean flow, and a brief description of the well-known algorithms and expressions [5, 23, 24] which can also be used by others to write their own library of MATLAB routines (functions) to compute the angular (periodic) Mathieu functions, and the more formidable radial (modified) Mathieu functions. Next, the procedure to compute the parametric zeros corresponding to the rigid-wall duct modes is presented which is followed by tables of non-dimensional resonance frequencies and the q parameters for ellipses ranging from nearly circular to extremely flattened. Based on tabulated data, the effect of aspect-ratio on the variation of non-dimensional resonance frequencies of radial, even and odd modes and their mode shapes is studied for the first few orders, and interpolating polynomial expressions are developed which may be used to accurately estimate resonance frequencies for any aspect-ratio. For the sake of completeness, the non-dimensional resonance frequencies of a clamped vibrating elliptical membrane which satisfies the soft (Dirichlet) boundary condition is also documented in Appendix for a range of aspect-ratio.

2.2 The Convective 3-D Helmholtz Equation in Elliptical Cylindrical Co-ordinates

The convective 3-D homogeneous wave equation in the Cartesian co-ordinates (x, y, z) is given by [42]

$$\frac{\partial^2 p}{\partial x^2} + \frac{\partial^2 p}{\partial y^2} + \left(1 - \frac{U_0^2}{c_0^2}\right)\frac{\partial^2 p}{\partial z^2} - 2\frac{U_0}{c_0^2}\frac{\partial^2 p}{\partial z \partial t} - \frac{1}{c_0^2}\frac{\partial^2 p}{\partial t^2} = 0, \qquad (2.1)$$

where $p = p(x, y, z, t)$ is the acoustic pressure field, U_0 is the subsonic uniform mean flow along the positive z direction, c_0 is the sound speed and t denotes time. The relationship between the Cartesian and elliptical cylindrical co-ordinates is defined according to the transformation $x + jy = \cosh(\xi + j\eta)$ which yields

$$x = h \cosh \xi \cos \eta,$$
$$y = h \sinh \xi \sin \eta,$$
$$z = z, \tag{2.2–2.4}$$

and the scale factors for elliptical co-ordinates is

$$h_\xi = h_\eta = \frac{h}{\sqrt{2}} (\cosh 2\xi - \cos 2\eta)^{1/2}, h_z = 1 \tag{2.5–2.7}$$

where $j = \sqrt{-1}$, $h = (D_1/2)e$ is the semi-interfocal distance, ξ and η are the radial and angular elliptical co-ordinates, see Chap. 9 of McLachlan [5]. Here, $e = \sqrt{1 - (D_2/D_1)^2}$ is the eccentricity (written in italics) of the ellipse, D_1 and D_2 are the major- and minor-axis of an elliptical cross-section, respectively, where $0 \le e \le 1$. From Eqs. (2.2) and (2.3), we get

$$\frac{x^2}{h^2 \cosh^2 \xi} + \frac{y^2}{h^2 \sinh^2 \xi} = \cos^2 \eta + \sin^2 \eta = 1 \tag{2.8a}$$

and

$$\frac{x^2}{h^2 \cos^2 \eta} - \frac{y^2}{h^2 \sin^2 \eta} = \cosh^2 \xi - \sinh^2 \xi = 1 \tag{2.8b}$$

Equation 2.8a represents a family of confocal ellipses with major-axis $D_1 = 2h \cosh \xi$, minor-axis $D_2 = 2h \sinh \xi$, the common foci being the points $x = \pm h$ and $y = 0$. Similarly, Eq. (2.8b) represents a family of confocal hyperbolae with the same common foci. The two families of confocal ellipses and hyperbolas intersect orthogonally where each intersection corresponds to a point (x, y) given by Eqs. (2.2) and (2.3). This is illustrated in Fig. 2.1 which shows an elliptical co-ordinate system wherein the points $\mathbf{F}$ and $\mathbf{F'}$ denote the two foci and the thick line $\mathbf{FF'}$ represents the interfocal axis on which $\xi = 0$ and $\eta = [0, \pi]$. At points on the x-axis lying outside the interfocal axis, the elliptical co-ordinate ξ ranges from $[0, \infty)$. In passing once around an ellipse (constant ξ), the angle η varies from 0 to 2π whereas on a hyperbola (constant η), ξ varies from zero at the interfocal axis to $\xi_0 = \cosh^{-1}(1/e)$ at the boundary of the bounding confocal ellipse. For large $\xi_0 \to \infty$, $e \to 0$ i.e., the ellipse tends to a circle of radius R_0. Since $h = (D_1/2)e$, in the limit $h \to 0$, the foci tend to coalesce at the origin, and $h \cosh \xi \to h \sinh \xi \to R_0$ However, if h is a constant, as $R_0 \to \infty, \xi \to \infty$ and $e \to 0$, so that the confocal ellipses

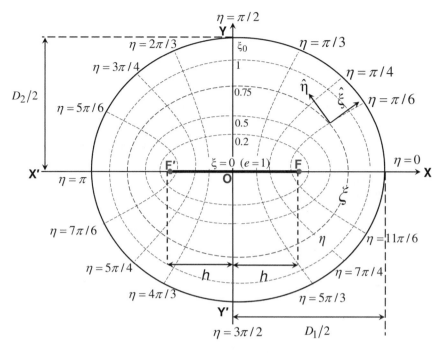

Fig. 2.1 An elliptical co-ordinate system. Here, F' and F denote the two focal points and the line joining them is the interfocal axis on which $\xi = 0$ ($e = 1$) and $\eta = [0, \pi]$.

tend to become concentric circles. Furthermore, using Eq. (2.8b), we see that as $h \rightarrow 0$, $y/x \rightarrow \pm \tan \eta$, so $\eta \rightarrow \phi$, $\cos \eta \rightarrow \cos \phi$, the confocal hyperbolae eventually become radii of the circle and make angles ϕ with the x-axis.

The relationship between the Cartesian and elliptical cylindrical co-ordinates shown in Eqs. (2.2–2.4) and (2.1) is used to derive the 3-D convective Helmholtz equation governing acoustic wave propagation in an infinitely long elliptical cylindrical waveguide (shown in Fig. 2.2) given by [42–44]

$$\left\{ \frac{2}{h^2(\cosh 2\xi - \cos 2\eta)} \left\{ \frac{\partial^2}{\partial \xi^2} + \frac{\partial^2}{\partial \eta^2} \right\} + (1 - M_0^2) \frac{\partial^2}{\partial z^2} - 2jM_0 k_0 \frac{\partial}{\partial z} + k_0^2 \right\} p = 0,$$

$$(2.9a)$$

or

$$\left\{ \frac{1}{h_\xi h_\eta} \left\{ \frac{\partial^2}{\partial \xi^2} + \frac{\partial^2}{\partial \eta^2} \right\} + (1 - M_0^2) \frac{\partial^2}{\partial z^2} - 2jM_0 k_0 \frac{\partial}{\partial z} + k_0^2 \right\} p = 0, \qquad (2.9b)$$

where $M_0 = U_0/c_0$ denotes the subsonic Mach number along the positive z direction. In Eq. (2.9), a time-harmonic field is assumed implying $p(\xi, \eta, z, t) = p(\xi, \eta, z)\mathbf{e}^{j\omega t}$

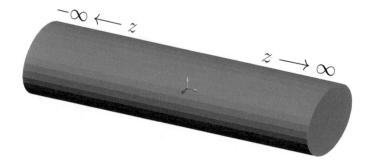

Fig. 2.2 An elliptical cylindrical waveguide extending infinitely in both directions and carrying a uniform mean flow toward the positive z direction

$k_0 = \omega/c_0$ is the forcing wavenumber corresponding to the excitation frequency ω and $\mathbf{e}^{j(\cdot)}$ denotes the complex exponential.

The acoustic pressure field $p(\xi, \eta, z)$ is of the form $p(\xi, \eta, z) = p_\xi(\xi) p_\eta(\eta) p_z(z)$, where one notices that the elliptical cylindrical co-ordinates allow for a separable solution in each of the three co-ordinates [35]. The acoustic pressure field $p_z(z)$ along the z direction satisfies

$$\left(1 - M_0^2\right) \frac{d^2 p_z}{dz^2} - 2 j M_0 k_0 \frac{\partial p_z}{\partial z} + k_z^2 p_z = 0 \tag{2.10}$$

where k_z is the axial wavenumber. The spatial variation of the acoustic pressure field along the transverse (elliptical) co-ordinates is given by

$$\frac{1}{p_\xi} \frac{\partial^2 p_\xi(\xi)}{\partial \xi^2} + \frac{1}{p_\eta} \frac{\partial^2 p_\eta(\eta)}{\partial \eta^2} + \frac{h^2}{2} (\cosh 2\xi - \cos 2\eta)\left(k_0^2 - k_z^2\right) = 0 \tag{2.11}$$

Substituting,

$$\frac{4q}{h^2} = k_0^2 - k_z^2 \tag{2.12}$$

in Eq. (2.11), and separating the wave equations along η and ξ directions yield

$$\frac{d^2 p_\eta}{d\eta^2} + (a - 2q \cos 2\eta) p_\eta = 0 \tag{2.13}$$

$$\frac{d^2 p_\xi}{d\xi^2} - (a - 2q \cosh 2\xi) p_\xi = 0 \tag{2.14}$$

where the parameter q plays the same role as the transverse or the radial wavenumber k_r in the case of a circular duct [5]. Furthermore, the separation constant a in Eqs. (2.13) and (2.14) is so chosen that solutions are periodic in η, so that $p_\eta(\eta) = p_\eta(\eta + 2\pi)$. In fact, it is the dependence on two parameter a and q that renders the computation of Mathieu functions far more involved than their Bessel function counterparts, perhaps, making these functions the most difficult ones in mathematical physics, see Arfken and Weber [45]. It is noted that in the limit $h \to 0$, the equations for the duct of elliptic cross-section reduce to those appropriate for the circular case [41].

Equations (2.10), (2.13) and (2.14), therefore, represent a set of ordinary differential equations that govern acoustic wave propagation along the axial or z, elliptical angular and radial directions, respectively. Here, we first present the well-known algorithms and expressions for computing the angular and radial elliptical solutions followed by tabulation and analysis of the non-dimensional resonance frequencies of rigid-wall transverse modes for different aspect-ratios.

2.3   Solution of Wave Equation in Angular and Radial Co-ordinates

2.3.1   Computation of Mathieu Functions

The angular component $p_\eta(\eta)$ of the acoustic pressure field in the elliptical chamber satisfies periodicity with period π or 2π Therefore, the solution of Eq. (2.13) is obtained as an infinite series of *cosine* or *sine* functions. To this end, it first noted that for a given q, there exists an infinite sequence of characteristic values $a = a_{\text{even}}(m)$, $m = 0, 1, 2, \ldots$ which corresponds to even periodic solution of Eq. (2.13) and another infinite sequence of characteristic values $a = a_{\text{odd}}(m)$, $m = 1, 2, 3, \ldots$ which corresponds to odd periodic solution. The characteristic value a is expanded about an integer m such that

$$a = m^2 + \alpha_1 q + \alpha_2 q^2 + \alpha_3 q^3 + \cdots \tag{2.15}$$

where $m = 0, 1, 2, \ldots$ for even solutions and $m = 1, 2, 3, \ldots$ for odd solutions, whereby it is noted in the limit $q \to 0$, $a \to m^2$ which corresponds to the case of circular duct. Since the solution of Eq. (2.13) is to reduce to, say, $\cos m\eta$ or $\sin m\eta$ when q vanishes, it is assumed that

$$p_\eta = \cos m\eta + q c_1(\eta) + q^2 c_2(\eta) + q^3 c_3(\eta) + \cdots, \tag{2.16}$$

or

$$p_\eta = \sin m\eta + q s_1(\eta) + q^2 s_2(\eta) + q^3 s_3(\eta) + \cdots \qquad (2.17)$$

On substituting the series solution given by Eqs. (2.16) into Eq. (2.13), collecting the coefficients of like powers of q, and equating them to zero, yields a set of simultaneous second-order ordinary differential equations (ODE) for functions $c_1(\eta), c_2(\eta), c_3(\eta)$. The particular integer (PI) of the ODEs represents the solution of these functions. Furthermore, since p_η must be periodic, the inhomogeneous secular terms in the ODE corresponding to q^i are set to zero which yields appropriate α_i and the PI solution of the simplified ODE yields an expression for $c_i(\eta), i = 1, 2, 3, \ldots$ Beginning from q^0 and proceeding up to q^n, one obtains the first n expansion terms of Eqs. (2.15) and (2.16). A similar procedure may also be followed for obtaining the $\sin m\eta$ series solution given by Eq. (1.17). Using this approach, McLachlan [5] computed the first several terms of $\cos(m\eta)$ and $\sin m\eta$ series as well as the characteristic values $a = a_{\text{even}}(m), a_{\text{odd}}(m)$ for first few orders $m = 0, 1, 2, \ldots$. However, these series or formulae are useful to compute p_η or a, only if q is small as the series tend to diverge for large q-values. Nevertheless, it is noted that regardless of m, the series expansion about $\cos m\eta$ consists of only *cosine* terms, while the series expansion about $\sin m\eta$ consists of only *sine* terms. For this reason, Eqs. (2.16) and (2.17) are often referred to as the *even* and *odd* series solution, respectively. It can be shown that for even values of order $m = 0, 2, 4, \ldots, 2n$ the even series solution consists of only $\cos 2\eta, \cos 4\eta, \cos 6\eta, \ldots$ terms, while for odd values $m = 1, 3, 5, \ldots, 2n - 1$. the even series solution consists of only $\cos \eta, \cos 3\eta, \cos 5\eta, \ldots$ terms. Similarly, for $m = 2, 4, \ldots, 2n$, the odd series solution consists of $\sin 2\eta, \sin 4\eta, \sin 6\eta, \ldots$ terms, while for $m = 1, 3, 5, \ldots, 2n - 1$, the odd series solution consists of $\sin \eta, \sin 3\eta, \sin 5\eta, \ldots$ terms. Therefore, for a given order m, there exists the following four categories of (periodic) solutions of Eq. (2.13), see Refs. [5, 23]:

$$p_\eta(\eta) = ce_{2n}(\eta, q) = \sum_{r=0}^{\infty} A_{2r}^{2n} \cos(2r\eta), \qquad m = 2n, n = 0, 1, 2, \ldots \quad (2.18)$$

$$p_\eta(\eta) = ce_{2n-1}(\eta, q) = \sum_{r=1}^{\infty} A_{2r-1}^{2n-1} \cos((2r - 1)\eta), \quad m = 2n - 1, n = 1, 2, 3, \ldots \ (2.19)$$

$$p_\eta(\eta) = se_{2n}(\eta, q) = \sum_{r=1}^{\infty} B_{2r}^{2n} \sin(2r\eta), \ldots \qquad m = 2n, n = 1, 2, 3, \qquad (2.20)$$

$$p_\eta(\eta) = se_{2n-1}(\eta, q) = \sum_{r=1}^{\infty} B_{2r-1}^{2n-1} \sin((2r - 1)\eta), \quad m = 2n - 1, n = 1, 2, 3, \ldots \ (2.21)$$

Following McLachlan's notation [5], the functions $ce_{2n}(\eta, q)$, $ce_{2n-1}(\eta, q)$, $se_{2n}(\eta, q)$ and $se_{2n-1}(\eta, q)$ are referred to as the angular Mathieu function (or simply, the Mathieu function) of the Even-even type, Even–odd type, Odd–even type and Odd-odd type, respectively. Equations (2.18–2.21) represent the trigonometric expansion of Mathieu functions (used for actual computation) wherein A_{2r}^{2n}, A_{2r-1}^{2n-1}, B_{2r}^{2n} and B_{2r-1}^{2n-1} are the expansion coefficients that depend on the value of the q parameter.

On inserting each of the Eqs. (2.18–2.21) into Eq. (2.13), one obtains an infinite set of recurrence relations between the coefficients of each type of solution shown as follows [5, 23].

$$a A_0^{2n} - q A_2^{2n} = 0,$$

Even-even
$$(a - 4) A_2^{2n} - q(2A_0^{2n} + A_4^{2n}) = 0,$$
$$\{a - (2k)^2\} A_{2k}^{2n} - q(A_{2k-2}^{2n} + A_{2k+2}^{2n}) = 0, \quad k = 2, 3, 4, \ldots$$
$$\tag{2.22}$$

Even-odd
$$(a - 1) A_1^{2n-1} - q(A_1^{2n-1} + A_3^{2n-1}) = 0,$$
$$\{a - (2k + 1)^2\} A_{2k+1}^{2n-1} - q(A_{2k-1}^{2n-1} + A_{2k+3}^{2n-1}) = 0, \quad k = 1, 2, 3, \ldots$$
$$\tag{2.23}$$

Odd-even
$$(a - 4) B_2^{2n} - q B_4^{2n} = 0,$$
$$\{a - (2k)^2\} B_{2k}^{2n} - q(B_{2k-2}^{2n} + B_{2k+2}^{2n}) = 0, \quad k = 2, 3, 4, \ldots$$
$$\tag{2.24}$$

Odd-odd
$$(a - 1) B_1^{2n+1} + q(B_1^{2n+1} - B_3^{2n+1}) = 0,$$
$$\{a - (2k + 1)^2\} B_{2k+1}^{2n-1} - q(B_{2k-1}^{2n-1} + B_{2k+3}^{2n-1}) = 0, \quad k = 1, 2, 3, \ldots$$
$$\tag{2.25}$$

Following Ref. [24], the expansion coefficients A_{2r}^{2n}, A_{2r-1}^{2n-1}, B_{2r}^{2n} and B_{2r-1}^{2n-1} are determined by recasting the above recurrence relationships into the matrix eigenvalue problem and then truncating the infinite matrix into a finite-size matrix whereby only the first several recurrence relations are considered. The matrix forms are shown as follows.

Even-even
$$
\begin{bmatrix}
-a & q & 0 & 0 & 0 & 0 & \ldots \\
2q & 2^2 - a & q & 0 & 0 & 0 & \ldots \\
0 & q & 4^2 - a & q & 0 & 0 & \ldots \\
0 & 0 & q & 6^2 - a & q & 0 & \ldots \\
\cdot & \cdot & \cdot & \cdot & \cdot & \cdot & \cdots \\
\cdot & \cdot & \cdot & \cdot & \cdot & \cdot & \cdots \\
\cdot & \cdot & \cdot & \cdot & \cdot & \cdot & \cdots
\end{bmatrix}
\begin{Bmatrix}
A_0^{2n} \\
A_2^{2n} \\
A_4^{2n} \\
A_6^{2n} \\
\cdot
\end{Bmatrix} = 0, \tag{2.26}
$$

Even-odd
$$
\begin{bmatrix}
1+q-a & q & 0 & 0 & 0\,0\ldots \\
q & 3^2-a & q & 0 & 0\,0\ldots \\
0 & q & 5^2-a & q & 0\,0\ldots \\
0 & 0 & q & 7^2-a & q\,0\ldots \\
\cdot & \cdot & \cdot & \cdot & \cdot\cdot\cdot\cdot\cdot \\
\cdot & \cdot & \cdot & \cdot & \cdot\cdot\cdot\cdot\cdot \\
\cdot & \cdot & \cdot & \cdot & \cdot\cdot\cdot\cdot\cdot
\end{bmatrix}
\begin{Bmatrix}
A_1^{2n-1} \\
A_3^{2n-1} \\
A_5^{2n-1} \\
A_7^{2n-1} \\
\cdot \\
\cdot
\end{Bmatrix} = 0, \quad (2.27)
$$

Odd-even
$$
\begin{bmatrix}
2^2-a & q & 0 & 0 & 0\,0\ldots \\
q & 4^2-a & q & 0 & 0\,0\ldots \\
0 & q & 6^2-a & q & 0\,0\ldots \\
0 & 0 & q & 8^2-a & q\,0\ldots \\
\cdot & \cdot & \cdot & \cdot & \cdot\cdot\cdot\cdot\cdot \\
\cdot & \cdot & \cdot & \cdot & \cdot\cdot\cdot\cdot\cdot \\
\cdot & \cdot & \cdot & \cdot & \cdot\cdot\cdot\cdot\cdot
\end{bmatrix}
\begin{Bmatrix}
B_2^{2n} \\
B_4^{2n} \\
B_6^{2n} \\
B_8^{2n} \\
\cdot \\
\cdot
\end{Bmatrix} = 0, \quad (2.28)
$$

Odd-odd
$$
\begin{bmatrix}
1-q-a & q & 0 & 0 & 0\,0\ldots \\
q & 3^2-a & q & 0 & 0\,0\ldots \\
0 & q & 5^2-a & q & 0\,0\ldots \\
0 & 0 & q & 7^2-a & q\,0\ldots \\
\cdot & \cdot & \cdot & \cdot & \cdot\cdot\cdot\cdot\cdot \\
\cdot & \cdot & \cdot & \cdot & \cdot\cdot\cdot\cdot\cdot \\
\cdot & \cdot & \cdot & \cdot & \cdot\cdot\cdot\cdot\cdot
\end{bmatrix}
\begin{Bmatrix}
B_1^{2n-1} \\
B_3^{2n-1} \\
B_5^{2n-1} \\
B_7^{2n-1} \\
\cdot \\
\cdot
\end{Bmatrix} = 0. \quad (2.29)
$$

The infinite matrices shown in Eqs. (2.26–2.29) are real, tridiagonal and symmetric for all the cases except for the case of Even-Even Mathieu functions, in which the matrix is slightly non-symmetric. The eigenvalue problem in Eqs. (2.18–2.21) is solved in MATLAB using the '*eig*' function whereby we obtain the eigenvector, i.e., the coefficients A_{2r}^{2n}, A_{2r-1}^{2n-1}, B_{2r}^{2n} and B_{2r-1}^{2n-1} as well as the eigenvalue a corresponding to each set of expansion coefficients. It may be mentioned here that in the a-q plane or the stability diagram, each eigenvalues loci corresponds to an even or an odd periodic solution. Furthermore, the region between two adjacent loci in the a-q plane corresponds to a non-periodic solution of Eq. (2.13) which may be growing (unstable) or decaying (stable). These regions, also known as *tongues* of stability and instability, occur alternatively and emanate from the confluence of even and odd eigenvalue loci corresponding to the same order m, see McLachlan [5].

The set of expansion coefficients (eigenvectors) for each a is further normalized subject to the following conditions [5]:

$$
\begin{aligned}
2\left(A_0^{2n}\right)^2 + \sum_{r=1}^{\infty}\left(A_{2r}^{2n}\right)^2 &= 1 \\
\sum_{r=1}^{\infty}\left(A_{2r-1}^{2n-1}\right)^2 = \sum_{r=1}^{\infty}\left(B_{2r}^{2n}\right)^2 = \sum_{r=1}^{\infty}\left(B_{2r-1}^{2n-1}\right)^2 &= 1.
\end{aligned}
\tag{2.30}
$$

The complete solution of Eq. (2.13) can, therefore, be expressed in the following form:

$$p_\eta(\eta) = A_c^1 ce_m(\eta, q) + A_s^1 se_m(\eta, q), \tag{2.31}$$

where the order $m = 0, 1, 2...$ for $ce_m(\eta, q_m)$, while $m = 1, 2, 3...$ for $se_m(\eta, q_m)$. As noted earlier, in the limit $q \to 0$, $a \to m^2$ which implies that the Mathieu functions reduce to trigonometric functions for the case of a circular duct, i.e., $ce_m(\eta, q_m) \to \cos m\theta$ and $se_m(\eta, q) \to \sin m\theta$.

2.3.2 Computation of Modified Mathieu Functions

Equation (2.14) can be obtained from Eq. (2.13) by replacing η with $j\xi$; therefore, its solution may also be obtained by introducing the same transformation in the angular Mathieu function solutions (Eqs. (2.18–2.21)) implying

$$p_\xi(\xi) = Ce_{2n}(\xi, q) = \sum_{r=0}^{\infty} A_{2r}^{2n} \cosh(2r\xi), \qquad m = 2n, n = 0, 1, 2, \ldots$$

$$p_\xi(\xi) = Ce_{2n-1}(\xi, q) = \sum_{r=1}^{\infty} A_{2r-1}^{2n-1} \cosh((2r-1)\xi), \quad m = 2n-1, n = 1, 2, 3, \ldots$$

$$p_\xi(\xi) = Se_{2n}(\xi, q) = \sum_{r=1}^{\infty} B_{2r}^{2n} \sinh(2r\xi), \qquad m = 2n, n = 1, 2, 3, \ldots$$

$$p_\xi(\xi) = Se_{2n-1}(\xi, q) = \sum_{r=1}^{\infty} B_{2r-1}^{2n-1} \sinh((2r-1)\xi), \quad m = 2n-1, n = 1, 2, 3, \ldots$$

$$\text{(2.32–2.35)}$$

Equations (2.32–2.35) describe the radial dependence of the acoustic pressure field in elliptical co-ordinates and are collectively termed as radial Mathieu functions or modified Mathieu functions. It is noted that Eqs. (2.32–2.35) have a similar form as Eqs. (2.18–2.21), respectively, the difference being that the trigonometric functions in the angular functions are replaced with hyperbolic functions in the radial functions. However, the expressions shown in Eqs. (2.32–2.35) are only 'formal' solutions of Eq. (2.14) and cannot be used for actual computations due to their extremely slow convergence unless ξ is small [5].

With a view to increase computational efficiency, the modified Mathieu functions are expressed as series of Bessel functions shown as follows [5, 23].

$$Ce_{2n}(\xi, q) = \frac{ce_{2n}(0, q)}{A_0^{2n}} \sum_{r=0}^{\infty} A_{2r}^{2n} J_{2r}(2\sqrt{q} \sinh \xi), \qquad\qquad m = 2n, n = 0, 1, 2, \ldots$$

$$Ce_{2n-1}(\xi, q) = \frac{ce'_{2n+1}(\pi/2, q)}{\sqrt{q} A_1^{2n-1}} \sum_{r=0}^{\infty} (-1)^{r+1} A_{2r+1}^{2n-1} J_{2r+1}(2\sqrt{q} \cosh \xi), \qquad m = 2n - 1, n = 1, 2, 3, \ldots$$

$$Se_{2n}(\xi, q) = \frac{se_{2n}(\pi/2, q)}{q B_2^{2n}} \tanh \xi \sum_{r=0}^{\infty} (-1)^{r+1} (2r + 2) \, B_{2r+2}^{2n} J_{2r+2}(2\sqrt{q} \cosh \xi), \quad m = 2n, n = 1, 2, 3, \ldots$$

$$Se_{2n-1}(\xi, q) = \frac{se'_{2n-1}(0, q)}{\sqrt{q} B_1^{2n-1}} \sum_{r=0}^{\infty} B_{2r+1}^{2n-1} J_{2r+1}(2\sqrt{q} \sinh \xi), \qquad\qquad m = 2n - 1, n = 1, 2, 3, \ldots$$

$$(2.36\text{--}2.39)$$

where J_r denotes the Bessel function of the first kind of positive integer order $r = 0, 1, 2, \ldots$ computed using the available Bessel function routines in MATLAB.

A more rapidly converging series, however, is that expressed in terms of products of Bessel functions given by [5, 23]

$$Ce_{2n}(\xi, q) = \sqrt{\frac{\pi}{2}} \frac{(-1)^n}{A_0^{2n}} \sum_{r=0}^{\infty} (-1)^r A_{2r}^{2n} J_r(v_1) J_r(v_2), \qquad\qquad m = 2n, n = 0, 1, 2, \ldots$$

$$Ce_{2n-1}(\xi, q) = \sqrt{\frac{\pi}{2}} \frac{(-1)^n}{A_1^{2n-1}} \sum_{r=0}^{\infty} (-1)^r A_{2r+1}^{2n-1} \{J_r(v_1) J_{r+1}(v_2) + J_r(v_2) J_{r+1}(v_1)\}, \quad m = 2n - 1, n = 1, 2, 3, \ldots$$

$$Se_{2n}(\xi, q) = \sqrt{\frac{\pi}{2}} \frac{(-1)^n}{B_2^{2n}} \sum_{r=1}^{\infty} (-1)^r B_{2r}^{2n} \{J_{r-1}(v_1) J_{r+1}(v_2) - J_{r-1}(v_2) J_{r+1}(v_1)\}, \quad m = 2n, n = 1, 2, 3, \ldots$$

$$Se_{2n-1}(\xi, q) = \sqrt{\frac{\pi}{2}} \frac{(-1)^n}{B_1^{2n-1}} \sum_{r=0}^{\infty} (-1)^r B_{2r+1}^{2n-1} \{J_r(v_1) J_{r+1}(v_2) - J_r(v_2) J_{r+1}(v_1)\}, \quad m = 2n - 1, n = 1, 2, 3, \ldots$$

$$(2.40\text{--}2.43)$$

where $v_1 = \sqrt{q} e^{-\xi}$ and $v_2 = \sqrt{q} e^{\xi}$. Similar to the angular Mathieu functions, we consider the first 50 terms of the series given by Eqs. (2.40–2.43) to compute the radial or modified Mathieu functions.

Following McLachlan's notation [5], these functions $Ce_{2n}(\xi, q)$, $Ce_{2n-1}(\xi, q)$, $Se_{2n}(\xi, q)$ and $Se_{2n-1}(\xi, q)$ are henceforth, referred to as the radial or modified Mathieu function of the Even-even type, Even–odd type, Odd–even type and Odd-odd type, respectively, same as their angular counterparts. Equations (2.40–2.43) are actually used for computing the transverse resonance frequencies of, and analysis of attenuation performance of elliptical chamber mufflers.

2.4 Acoustic Pressure Field in a Rigid-Wall Elliptical Cylindrical Waveguide

The acoustic field in an elliptical cylindrical waveguide must satisfy the following conditions [35]:

(1) Since $p(\xi, \eta)$ is single-valued, it must be periodic in η, with a period at most 2π,

(2) $p(\xi, \eta)$ is a continuous function; in particular, it is continuous across the interfocal line, so that

$$p(\xi = 0, \eta) = p(\xi = 0, -\eta). \tag{2.44}$$

(3) Also, on crossing the interfocal line, the continuity of pressure gradient must be satisfied, so that

$$\frac{\partial}{\partial \xi} p(\xi = 0, \eta) = -\frac{\partial}{\partial \xi} p(\xi = 0, -\eta), \tag{2.45}$$

which signifies that the normal acoustic velocity at the interfocal axis must be continuous.

The angular Mathieu functions ce_m, se_m (of order m) given by Eqs. (2.18–2.21) satisfy the periodicity requirement set out by condition (1). The conditions (2) and (3) further imply that the acoustic field at an elliptical cross-section is given by

$$p(\xi, \eta, z, t) = \left\{ \begin{array}{l} \sum\limits_{m=0,1,2,\ldots}^{\infty} Ce_m(\xi, q)ce_m(\eta, q)\left(C_m^1 \mathbf{e}^{-jk_{z,m}^+ z} + C_m^2 \mathbf{e}^{jk_{z,m}^- z}\right) + \\ \sum\limits_{m=1,2,\ldots}^{\infty} Se_m(\xi, q)se_m(\eta, q)\left(S_m^1 \mathbf{e}^{-jk_{z,m}^+ z} + S_m^2 \mathbf{e}^{jk_{z,m}^- z}\right) \end{array} \right\} \mathbf{e}^{j\omega t}, \tag{2.46}$$

where

$$k_{z,m}^{\pm} = \frac{\mp M_0 k_0 + \sqrt{k_0^2 - \dfrac{4q_m}{h^2}\left(1 - M_0^2\right)}}{\left(1 - M_0^2\right)}, \tag{2.47a, b}$$

and $\mathbf{e}^{j(\cdot)}$ denotes the complex exponential shown in an upright bold manner. In addition to aforementioned conditions (1–3), at the boundary $\xi = \xi_0 = \cosh^{-1}(1/e)$ of the *rigid-wall* or *hard-wall* elliptical waveguide, the normal component of the acoustic pressure gradient must be zero, i.e., the acoustic particle velocity u_ξ along the ξ direction should be zero. Therefore,

$$u_{\xi=\xi_0} = -\frac{1}{jk_0\rho_0c_0}\frac{1}{h_\xi}\frac{\partial p_\xi}{\partial \xi}\bigg|_{\xi=\xi_0} = 0. \tag{2.48}$$

which yields (the Neumann boundary condition)

$$\frac{d}{d\xi}Ce_m\left(\xi, q_{m,n}\right)\bigg|_{\xi=\xi_0} = 0, \quad m = 0, 1, \ldots, \infty, \quad n = 1, 2, \ldots, \infty \tag{2.49}$$

$$\frac{d}{d\xi}Se_m\left(\xi, \overline{q}_{m,n}\right)\bigg|_{\xi=\xi_0} = 0, \quad m = 1, 2, \ldots, \infty, \quad n = 1, 2, \ldots, \infty \tag{2.50}$$

Equations (2.49) and (2.50) imply that there exists an infinite number of parametric zeros $q_{m,n}$ or $\overline{q}_{m,n}$ of the derivative of the even/odd modified Mathieu function for a given order m. These parametric zeros are termed as the 'q' parameters for a rigid-wall elliptical duct. The corresponding non-dimensional resonance frequencies of transverse acoustic modes in a waveguide with a stationary medium are shown as follows.

$$\left(k_0\frac{D_1}{2}\right)\bigg|_{(m,n)e} = \left(\frac{2\sqrt{q_{m,n}}}{e}\right), \quad m = 0, 1, \ldots, \infty, n = 1, 2, \ldots, \infty, \tag{2.51}$$

and

$$\left(k_0\frac{D_1}{2}\right)\bigg|_{(m,n)o} = \left(\frac{2\sqrt{\overline{q}_{m,n}}}{e}\right), m = 1, 2, \ldots, \infty, n = 1, 2, \ldots, \infty. \tag{2.52}$$

It is important to clarify here the subscript e in Eq. (2.51) denotes the even modes while the subscript o in Eq. (2.52) denotes the odd modes—this convention is followed throughout this book. For a rigid-wall infinite elliptical cylindrical waveguide carrying a uniform subsonic mean flow (toward the positive z direction), the three-dimensional (3-D) acoustic pressure field is expressed as infinite modal summation given by

$$p(\xi, \eta, z, t)$$
$$= \left\{ \begin{array}{l} \sum_{m=0,1,2\ldots}^{\infty}\sum_{n=1,2\ldots}^{\infty} Ce_m\left(\xi, q_{m,n}\right)ce_m\left(\eta, q_{m,n}\right)\left(C^1_{m,n}e^{-jk^+_{z,m,n}z} + C^2_{m,n}e^{jk^-_{z,m,n}z}\right) \\ + \sum_{m=1,2\ldots}^{\infty}\sum_{n=1,2\ldots}^{\infty} Se_m\left(\xi, \overline{q}_{m,n}\right)se_m\left(\eta, \overline{q}_{m,n}\right)\left(S^1_{m,n}e^{-j\overline{k}^+_{z,m,n}z} + S^2_{m,n}e^{j\overline{k}^-_{z,m,n}z}\right) \end{array} \right\}e^{j\omega t}, \tag{2.53}$$

where C's and S's are arbitrary constants and the convective axial wave numbers $k^\pm_{z,m,n}$ and $\overline{k}^\pm_{z,m,n}$ are given by

$$k_{z,m,n}^{\pm} = \frac{\mp M_0 k_0 + \sqrt{k_0^2 - \dfrac{4q_{m,n}}{h^2}\left(1 - M_0^2\right)}}{\left(1 - M_0^2\right)}, \qquad (2.54\mathrm{a,\,b})$$

$$\overline{k}_{z,m,n}^{\pm} = \frac{\mp M_0 k_0 + \sqrt{k_0^2 - \dfrac{4\overline{q}_{m,n}}{h^2}\left(1 - M_0^2\right)}}{\left(1 - M_0^2\right)}, \qquad (2.55\mathrm{a,\,b})$$

From Eqs. (2.54) and (2.55), it is evident that an even or odd mode (m, n) will propagate or will become cut-on if

$$\left(k_0 \frac{D_1}{2}\right)\Bigg|_{(m,n)e} \geq \left(\frac{2\sqrt{q_{m,n}}}{e}\right)\sqrt{1 - M_0^2}, \quad m = 0, 1, \ldots, \infty, n = 1, 2, \ldots, \infty,$$

$$(2.56)$$

and

$$\left(k_0 \frac{D_1}{2}\right)\Bigg|_{(m,n)o} \geq \left(\frac{2\sqrt{\overline{q}_{m,n}}}{e}\right)\sqrt{1 - M_0^2}, m = 1, 2, \ldots, \infty, n = 1, 2, \ldots, \infty,$$

$$(2.57)$$

else if the excitation frequency is less than that given by Eqs. (2.51) or (2.52), the corresponding even or odd (m, n) mode will be cut-off. In other words, such a transverse mode does not propagate but decays exponentially and is called an evanescent mode. Therefore, as expected, the uniform mean flow results in lowering of the cut-on frequency, see Munjal [42].

For $m \neq 0, n = 1, 2, 3, \ldots$ $Ce_m\left(\xi, q_{m,n}\right)ce_m\left(\eta, q_{m,n}\right)$ and $Se_m\left(\xi, \overline{q}_{m,n}\right)se_m\left(\eta, \overline{q}_{m,n}\right)$ modes are referred to as the even and odd transverse acoustic modes, respectively, of an elliptical waveguide. The even and odd modes are henceforth denoted as $(m, n)e$ and $(m, n)o$, respectively. It is important to note from Eqs. (2.53) and (2.56) that $Ce_0\left(\xi, q_{0,1}\right)ce_0\left(\eta, q_{0,1}\right)$ mode $(m = 0, n = 1)$ denotes the plane wave mode for which the acoustic pressure is constant throughout the elliptical cross-section for this mode. The plane wave mode is the fundamental mode and propagates regardless of the excitation frequency because $q = 0$ is always the first parametric zero (or root) of Eq. (2.49) for $m = 0$, i.e., $q_{0,1} = 0$ for any aspect-ratio D_2/D_1 as will be observed in the ensuing tables. Furthermore, for $m = 0, n = 2, 3, 4, \ldots, Ce_0\left(\xi, q_{0,n}\right)ce_0\left(\eta, q_{0,n}\right)$ modes are referred to as the radial modes.

In the limiting case of the elliptical cross-section tending to a circular cross-section of diameter D_0, i.e., when $D_2/D_1 \to 1$ or $e \to 0$, $Ce_m(\cdot) \to Se_m(\cdot) \to J_m(\cdot)$, $ce_m(\cdot) \to \cos(m\theta)$ and $se_m(\cdot) \to \sin(m\theta)$ Therefore, Eq. (2.53) transitions to the well-known form of the acoustic pressure field in a rigid-wall circular cylindrical waveguide [42]

$p(r, \theta, z, t)$

$$= \left\{ \sum_{m=0,1,2\dots}^{\infty} \sum_{n=0,1,2\dots}^{\infty} J_m\left(\alpha_{mn} \frac{r}{R_0}\right) \left(A_{m,n}^1 \cos m\theta + A_{m,n}^2 \sin m\theta\right) \left(C_{m,n}^1 e^{-jk_{z,m,n}^+ z} + C_{m,n}^2 e^{jk_{z,m,n}^- z}\right) \right\} e^{j\omega t},$$

$$(2.58)$$

where r and θ denote radial and azimuthal co-ordinates, respectively, $R_0 = D_0/2$ is the radius of chamber and A's are arbitrary constants. In Eq. (2.58), $\alpha_{m,n}$ is the non-dimensional resonance frequency of the (m, n) mode and is obtained by

$$\alpha_{m,n} = k_{m,n} R_0 = \left. \frac{dJ_m\left(k_{m,n} r\right)}{d\left(k_{m,n} r\right)} \right|_{r=R_0} = \frac{1}{2} \left\{ J_{m-1}\left(\alpha_{m,n}\right) - J_{m+1}\left(\alpha_{m,n}\right) \right\} = 0 \quad (2.59)$$

where $k_{m,n}$ denotes the radial wavenumber and

$$k_{z,m,n}^{\pm} = \frac{\mp M_0 k_0 + \sqrt{k_0^2 - k_{m,n}^2\left(1 - M_0^2\right)}}{\left(1 - M_0^2\right)}. \quad (2.60)$$

It is noted that for $(0, 0)$ mode, $\alpha_{0,0} = 0$ which implies that this mode is always cut-on regardless of the excitation frequency. Further, since $J_0(\alpha_{0,0} = 0) = 1$, i.e., constant over the circular cross-section, the $(0, 0)$ mode of a rigid-wall circular waveguide pertains to the plane wave mode in the modal summation solution shown in Eq. (2.58).

At this stage, it is important to clarify a difference in the terminologies used for denoting the transverse modes of the rigid-wall elliptical and the circular waveguides. Firstly, the plane wave in an elliptical waveguide is given by the $(0, 1)$ mode, whereas the $(0, 0)$ mode represents the plane wave in a circular waveguide. Next, the $(0, 2)$, $(0, 3)$, ... $(0, n)$ modes of an elliptical waveguide represent the 1st, 2nd, ..., $(n-1)$th radial mode, respectively, whereas in a circular waveguide, $(0, 1)$, $(0, 2)$, ... $(0, n-1)$ modes denote the 1st, 2nd, ..., $(n-1)$th radial mode, respectively.

In general, the $Ce_m\left(\xi, q_{m,n}\right) ce_m\left(\eta, q_{m,n}\right)$ and $Se_m\left(\xi, \overline{q}_{m,n}\right) se_m\left(\eta, \overline{q}_{m,n}\right)$ modes of the ellipse are the counterparts of the $J_m\left(\alpha_{m(n-1)} \frac{r}{R_0}\right) \cos m\theta$ and $J_m\left(\alpha_{m(n-1)} \frac{r}{R_0}\right) \sin m\theta$ modes, respectively, of the circular waveguide, where $n = 1, 2, 3, \dots$.

On a related note, it is worth mentioning that the displacement field $\zeta(\xi, \eta, t)$ for free-vibrations of an elliptical membrane which is clamped (fixed) on all sides, i.e., has zero displacement at the elliptical boundary $\xi = \xi_0$, has the same form as Eq. (2.53) and is given by [5]

$$\zeta(\xi, \eta) = \left\{ \begin{array}{l} \displaystyle\sum_{m=0,1,2\ldots}^{\infty} \sum_{n=1,2,\ldots}^{\infty} C_{m,n} Ce_m\left(\xi, q_{m,n}\right) ce_m\left(\eta, q_{m,n}\right) \cos\left(\omega_{m,n}t + \varepsilon_{m,n}\right) \\[2em] \displaystyle\sum_{m=1,2\ldots}^{\infty} \sum_{n=1,2\ldots}^{\infty} S_{m,n} Se_m\left(\xi, \overline{q}_{m,n}\right) se_m\left(\eta, \overline{q}_{m,n}\right) \cos\left(\overline{\omega}_{m,n}t + \overline{\varepsilon}_{m,n}\right) \end{array} \right\},$$

$$(2.61)$$

where $q_{m,n}$ and $\overline{q}_{m,n}$ are the parametric zeros corresponding to the clamped boundary condition $\zeta(\xi = \xi_0, \eta) = 0$, and they may be obtained by numerically solving

$$Ce_m\left(\xi, q_{m,n}\right)\big|_{\xi=\xi_0} = 0, \quad m = 0, 1, \ldots, \infty, n = 1, 2, \ldots, \infty \qquad (2.62)$$

and

$$Se_m\left(\xi, \overline{q}_{m,n}\right)\big|_{\xi=\xi_0} = 0, \quad m = 1, 2, \ldots, \infty, n = 1, 2, \ldots, \infty. \qquad (2.63)$$

signifying the application of the Dirichlet boundary condition. In Eq. (2.61), $\omega_{m,n}$ and $\overline{\omega}_{m,n}$ are the angular eigenfrequency of the (m, n) even and odd modes, respectively. They are related to the parametric zeros as

$$\omega_{m,n} = c_0 \left(\frac{2\sqrt{q_{m,n}}}{h}\right), \qquad (2.64)$$

$$\overline{\omega}_{m,n} = c_0 \left(\frac{2\sqrt{\overline{q}_{m,n}}}{h}\right), \qquad (2.65)$$

respectively, while $\varepsilon_{m,n}$ and $\overline{\varepsilon}_{m,n}$ are relative phases, $c_0 = \sqrt{T/m}$ is the wave propagation speed, T and m are the tension and mass per unit area of the membrane, respectively. The non-dimensional resonance frequencies of a clamped membrane can again be computed using Eqs. (2.51) and (2.52), for even and odd modes, respectively. Table 2.29 in the Appendix of this chapter presents the non-dimensional resonance frequencies of the clamped elliptical membrane for a range of aspect-ratio $D_2/D_1 = 0.1, 0.2(0.1), 0.9, 1.0$ corresponding to the *radial* and *even* circumferential/cross-modes, while Table 2.30 presents the non-dimensional resonance frequencies of *odd* circumferential/cross-modes. However, the focus of this chapter is the tabulation of the resonance frequencies of a rigid-wall elliptical acoustic waveguide (Sect. 2.6) which will be used in the analysis and design of reactive (unlined) short elliptical mufflers in the ensuing chapters.

Equation (2.61) also represents the free-oscillations of water (small displacements) in an elliptical lake as well as the magnetic wave H_z component of the electrodynamic waves along the axial or z direction in an elliptical waveguide. However, in these cases, $q_{m,n}$ and $\overline{q}_{m,n}$ are obtained by solving Eqs. (2.49) and (2.50) because the normal velocity of water and the tangential component E_η of the electric field are zero at the elliptical boundaries [5, 45].

2.5 Numerical Computation of the q Parameters

The procedure to compute the parametric zeros $q_{m,n}$ and $\overline{q}_{m,n}$ of the derivative of the even and odd radial Mathieu functions, respectively, based on the root-bracketing and bisection techniques is briefly described, see Ref. [35]. To this end, the normalized expansion coefficients A_{2r}^{2n}, A_{2r-1}^{2n-1}, B_{2r}^{2n} and B_{2r-1}^{2n-1} are first computed by solving the matrix eigenvalue problem (Sect. 2.3.1) for a given value of q, order m and even/odd function type. The expansion coefficients and Eqs. (2.40–2.43) are then used to compute the derivative of the radial Mathieu function. We seek real and positive parametric zeros; therefore, the derivative of radial Mathieu functions is computed for a range of discrete positive value starting from $q = 0$ for a given aspect-ratio value. For a given value of q parameter, the first 50 terms of the even-even, even–odd, odd–even and odd-odd series were considered in Eqs. (2.18–2.21), respectively, which was found sufficient to ensure a convergent solution set of expansion coefficient vectors and Mathieu functions for the maximum q parameter and order m of interest. Figure 2.3a–c shows a graphical variation of

$$\frac{\mathrm{d}}{\mathrm{d}\xi} Ce_m\left(\xi, q_{m,n}\right)\Bigg|_{\xi=\xi_0}, \tag{2.66}$$

versus q for $m = 0$ and $D_2/D_1 = (0.3, 0.6, 0.9)$, respectively.

It is observed from Figs. 2.3a that for a small aspect-ratio (eccentric ellipse), the graphs oscillate over a larger q range, and therefore, the parametric zeros are more widely spaced. On the other hand, Fig. 2.3c shows that for an ellipse with aspect-ratio close to unity (tending to a circle), oscillations are more rapid which signifies that the parametric zeros are located more closely or over a smaller interval. Since Eq. (2.66) is continuous, root-bracketing is used to detect its change of sign which gives the upper and lower bounds to locate a parametric zero. Further, in order to ensure that all parametric zeros are sequentially bracketed, Eq. (2.66) is evaluated at sufficiently fine intervals Δq, keeping in-mind the aspect-ratio for which the parametric zero needs to be evaluated. For instance, in Fig. 2.3a, $Ce_0(0.3095, q_{\mathrm{low}} = 27.25) = -0.008$ and $Ce_0(0.3095, q_{up} = 27.3) = 0.001$ brackets the parametric zero $q_{0,2} = 27.2932$, following which the bisection method [46] is then used to numerically evaluate this zero. The same procedure is followed for evaluating the larger parametric zeros, all of which are indicated by a '+' mark in Fig. 2.3a. It is noted that the q parameters in Figs. 2.3a–c are shown in Tables 2.7, 2.13 and 2.19, respectively, in the column corresponding to $m = 0$.

2.6 Tabulation of Parametric Zeros and Non-Dimensional Resonance Frequencies

The parametric zeros $q_{m,n}, \overline{q}_{m,n}$ and corresponding non-dimensional resonance frequency $0.5 k_{m,n} D_1, 0.5 \overline{k}_{m,n} D_1$ of a rigid-wall elliptical cylindrical waveguide for different values of the aspect-ratio

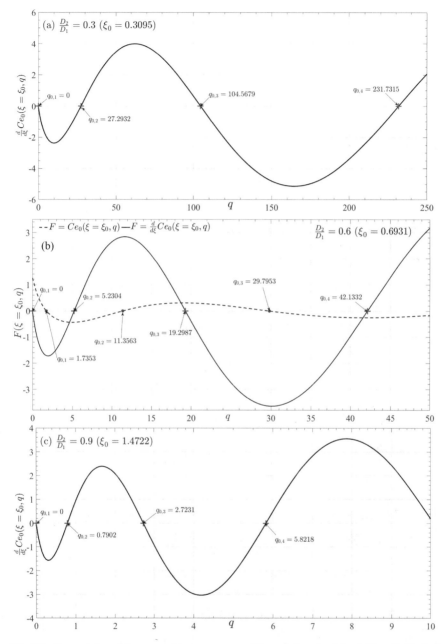

Fig. 2.3 Computation of the parametric zero $q_{m=0,n}$ of the radial mode of the rigid-wall elliptical waveguide for aspect-ratio (a) $D_2/D_1 = 0.3$, (b) $D_2/D_1 = 0.6$, (c) $D_2/D_1 = 0.9$ using the root-bracketing and bisection technique. The parametric zeros $q_{0,n=1,2,3...}$ are indicated by a $+$ mark. Part **b** also shows interlacing of the parametric zeros $q_{0,n}$ corresponding to the Dirichlet (clamped membrane) and Neumann (rigid-wall waveguide) boundary conditions

$$\frac{D_2}{D_1} = 0.01, 0.05, 0.1, 0.15, (0.05), \ldots, 0.95, (0.01), \ldots, 0.99, 0.999, \quad (2.67)$$

evaluated by solving Eqs. (2.49) and (2.50) are shown in Tables 2.1, 2.2, 2.3, 2.4, 2.5, 2.6, 2.7, 2.8, 2.9, 2.10, 2.11, 2.12, 2.13, 2.14, 2.15, 2.16, 2.17, 2.18, 2.19, 2.20, 2.21, 2.22, 2.23, 2.24, 2.25 up to four decimal places. The value in brackets () shown in the bold-font beneath each of the parametric zero in Tables 2.1, 2.2, 2.3, 2.4, 2.5, 2.6, 2.7, 2.8, 2.9, 2.10, 2.11, 2.12, 2.13, 2.14, 2.15, 2.16, 2.17, 2.18, 2.19, 2.20, 2.21, 2.22, 2.23, 2.24, 2.25 represents the non-dimensional transverse resonance (cut-on) frequency computed using Eqs. (2.51) and (2.52). *Note here that the major-axis D_1 is the non-dimensionalization variable; in other words, it is kept constant in the ensuing Tables.* Each table presents the first four zeros ($n = 1, 2, 3, 4$) of the radial mode $Ce_0(\xi, q_{0,n})ce_0(\eta, q_{0,n})$, the even modes $Ce_m(\xi, q_{m,n})ce_m(\eta, q_{m,n})$ and the odd modes $Se_m(\xi, q_{m,n})se_m(\eta, q_{m,n})$ shown in red, blue and black colors, respectively. Note here that the first seven and the first eight orders of the even and odd modes, respectively, are considered. Table 2.26 shows the non-dimensional cut-on frequencies of a circular cylindrical waveguide, i.e., $D_2/D_1 = 1$ evaluated by numerically solving Eq. (2.59) for the first nine orders $m = 0, 1, 2, \ldots, 8$ corresponding to the first nine roots $n = 0, 1, 2, \ldots, 8$ of each order up to four places of decimal.

To aid in interpretation of the tabulated values, Fig. 2.4 shows the shapes of members of the family of ellipses of *constant major-axis* depicting a gradual transition from a highly eccentric section of aspect-ratio 0.1 to a zero eccentric ellipse (perfectly circular).

Table 2.1 The q parameters and cut-on frequencies for aspect-ratio $D_2/D_1 = 0.01$ ($e = 0.9999$, $\xi_0 = 0.0100$)

Even Modes $\left.\frac{d}{d\xi}Ce_m(\xi, q_{m,n})\right|_{\xi=\xi_0} = 0$	$m = 0$	$m = 1$	$m = 2$	$m = 3$	$m = 4$	$m = 5$	$m = 6$	$m = 7$
$n = 1$	0 (0)	0.8897 (1.8866)	3.0387 (3.4866)	6.4250 (5.0698)	11.0466 (6.6476)	16.9026 (8.2230)	23.9926 (9.7969)	32.3166 (11.3701)
$n = 2$	24750.0136 (314.6585)	24907.8271 (315.6601)	25066.3914 (316.6632)	25225.6675 (317.6677)	25385.6737 (318.6736)	25546.0356 (319.6785)	25707.0816 (320.6846)	25867.1238 (321.6813)
$n = 3$	98842.9004 (628.8173)	99154.4976 (629.8076)	99466.5190 (630.7978)	99755.6043 (631.7138)	100058.9952 (632.6737)	100308.3815 (633.4616)	100578.9462 (634.3153)	100851.2663 (635.1735)
$n = 4$	222275.3012 (942.96878)	222711.1016 (943.8927)	223159.8047 (944.8431)	223455.6320 (945.4692)	223921.7899 (946.4548)	224226.2201 (947.0980)	224850.9263 (948.4164)	225877.1229 (950.5781)
Odd Modes $\left.\frac{d}{d\xi}Se_m(\xi, \bar{q}_{m,n})\right|_{\xi=\xi_0} = 0$	$m = 1$	$m = 2$	$m = 3$	$m = 4$	$m = 5$	$m = 6$	$m = 7$	$m = 8$
$n = 1$	6207.0887 (157.5780)	6286.3711 (158.5812)	6366.4087 (159.5875)	6447.2052 (160.5970)	6528.7639 (161.6096)	6611.0884 (162.6253)	6694.1823 (163.6441)	6778.0489 (164.6660)
$n = 2$	55628.7101 (471.7384)	55865.0596 (472.7394)	56102.1608 (473.7416)	56340.0149 (474.7448)	56578.6231 (475.7490)	56817.9866 (476.7543)	57058.1066 (477.7607)	57298.9843 (478.7681)
$n = 3$	154393.4179 (785.8979)	154786.8341 (786.8985)	155181.0012 (787.8998)	155575.9200 (788.9017)	155971.5912 (789.9043)	156368.0135 (790.9075)	156765.1906 (791.9113)	157163.0993 (792.9157)
$n = 4$	302501.2128 (1100.0572)	303051.6924 (1101.0577)	303602.9237 (1102.0586)	304154.8462 (1103.0599)	304707.5422 (1104.0616)	305260.3644 (1105.0627)	305814.1779 (1106.0647)	306364.5790 (1107.0596)

Table 2.2 The q parameters and cut-on frequencies for aspect-ratio $D_2/D_1 = 0.05$ ($e = 0.9987$, $\xi_0 = 0.0500$)

Even Modes $\left.\dfrac{\mathrm{d}}{\mathrm{d}\xi}Ce_m\left(\xi,q_{m,n}\right)\right|_{\xi=\xi_0}=0$	$m=0$	$m=1$	$m=2$	$m=3$	$m=4$	$m=5$	$m=6$	$m=7$
$n=1$	0 (0)	0.8875 (1.8865)	3.0303 (3.4859)	6.4064 (5.0685)	11.0134 (6.6456)	16.8503 (8.2201)	23.9168 (9.7932)	32.2126 (11.3654)
$n=2$	1000.1027 (63.3280)	1032.1994 (64.3362)	1065.0577 (65.3522)	1098.6862 (66.3759)	1133.0935 (67.4072)	1168.2880 (68.4461)	1204.2780 (69.4923)	1241.0717 (70.5459)
$n=3$	3969.2587 (126.1619)	4032.7071 (127.1663)	4096.9106 (128.1745)	4161.8737 (129.1868)	4227.6007 (130.2029)	4294.0960 (131.2228)	4361.3638 (132.2467)	4429.4085 (133.2743)
$n=4$	8907.3997 (188.9944)	9002.1989 (189.9975)	9097.7511 (191.0032)	9194.0592 (192.0115)	9291.1261 (193.0224)	9388.9548 (194.0359)	9487.5482 (195.0521)	9586.9092 (196.0708)
Odd Modes $\left.\dfrac{\mathrm{d}}{\mathrm{d}\xi}Se_m\left(\xi,\bar{q}_{m,n}\right)\right|_{\xi=\xi_0}=0$	$m=1$	$m=2$	$m=3$	$m=4$	$m=5$	$m=6$	$m=7$	$m=8$
$n=1$	253.8918 (31.9079)	270.3105 (32.9234)	287.5029 (33.9543)	305.4859 (35.0001)	324.2757 (36.0604)	343.8882 (37.1349)	364.3387 (38.3247)	385.6419 (39.2231)
$n=2$	2238.5577 (94.7453)	2286.3305 (95.7509)	2334.8606 (96.7618)	2384.1538 (97.7779)	2434.2160 (98.7991)	2485.0528 (99.8255)	2536.6699 (100.8933)	2589.0730 (101.8569)
$n=3$	6192.2060 (157.5782)	6271.3298 (158.5818)	6351.2075 (159.5885)	6431.8426 (160.5984)	6513.2385 (161.6114)	6595.3988 (162.6275)	6678.3270 (163.6690)	6762.0264 (164.6467)
$n=4$	12114.8398 (220.4106)	12225.3144 (221.4132)	12336.5413 (222.4182)	12448.5230 (223.4254)	12561.2622 (224.4348)	12674.7612 (225.4465)	12789.0227 (226.4765)	12904.0490 (227.4604)

Table 2.3 The q parameters and cut-on frequencies for aspect-ratio $D_2/D_1 = 0.1$ ($e = 0.9950$, $\xi_0 = 0.1003$)

Even Modes $\left.\dfrac{\mathrm{d}}{\mathrm{d}\xi}Ce_m\left(\xi,q_{m,n}\right)\right|_{\xi=\xi_0}=0$	$m=0$	$m=1$	$m=2$	$m=3$	$m=4$	$m=5$	$m=6$	$m=7$
$n=1$	0 (0)	0.8804 (1.8861)	3.0041 (3.4840)	6.3482 (5.0645)	10.9097 (6.6393)	16.6874 (8.2112)	23.6805 (9.7815)	31.8884 (11.3509)
$n=2$	251.9928 (31.9085)	268.3089 (32.9253)	285.3941 (33.9574)	303.2651 (35.0045)	321.9381 (36.0661)	341.4287 (37.1418)	361.7521 (38.2312)	382.9228 (39.3340)
$n=3$	992.6028 (63.3286)	1024.4986 (64.3381)	1057.1515 (65.3553)	1090.5699 (66.3803)	1124.7626 (67.4129)	1159.7376 (68.4530)	1195.5034 (69.5005)	1232.0680 (70.5533)
$n=4$	2221.7559 (94.7459)	2269.2294 (95.7528)	2317.4558 (96.7650)	2366.4408 (97.7823)	2416.1901 (98.8048)	2466.7095 (99.8324)	2518.0045 (100.8650)	2570.0808 (101.9027)
Odd Modes $\left.\dfrac{\mathrm{d}}{\mathrm{d}\xi}Se_m\left(\xi,\bar{q}_{m,n}\right)\right|_{\xi=\xi_0}=0$	$m=1$	$m=2$	$m=3$	$m=4$	$m=5$	$m=6$	$m=7$	$m=8$
$n=1$	64.8865 (16.1916)	73.4066 (17.2219)	82.7172 (18.2814)	92.8494 (19.3688)	103.8322 (20.4823)	115.6925 (21.6205)	128.4553 (22.7818)	142.1436 (23.9649)
$n=2$	561.2302 (47.6193)	585.3367 (48.6312)	610.2043 (49.6535)	635.8445 (50.6860)	662.2683 (51.7284)	689.4867 (52.7807)	717.5103 (53.8427)	746.3495 (54.9141)
$n=3$	1546.1113 (79.0374)	1585.7961 (80.0453)	1626.2353 (81.0595)	1667.4360 (82.0799)	1709.4051 (83.1065)	1752.1491 (84.1391)	1795.6749 (85.1778)	1839.9891 (86.2224)
$n=4$	3019.5368 (110.4543)	3074.7990 (111.4604)	3130.8128 (112.4711)	3187.5832 (113.4862)	3245.1151 (114.5058)	3303.4134 (115.5298)	3362.4831 (116.5581)	3422.3288 (117.5908)

Table 2.4 The q parameters and cut-on frequencies for aspect-ratio $D_2/D_1 = 0.15$ ($e = 0.9887$, $\xi_0 = 0.1511$)

Even Modes $\left.\dfrac{d}{d\xi}Ce_m\left(\xi,q_{m,n}\right)\right\|_{\xi=\xi_0}=0$	$m=0$	$m=1$	$m=2$	$m=3$	$m=4$	$m=5$	$m=6$	$m=7$
$n=1$	0 (0)	0.8687 (1.8854)	2.9606 (3.4807)	6.2516 (5.0579)	10.7377 (6.6287)	16.4171 (8.1963)	23.2884 (9.7621)	31.3507 (11.3265)
$n=2$	112.2617 (21.4332)	123.2655 (22.4591)	135.0419 (23.5075)	147.6146 (24.5774)	161.0062 (25.6681)	175.2383 (26.7785)	190.3311 (27.9079)	206.3039 (29.0553)
$n=3$	438.9891 (42.3837)	460.2728 (43.3990)	482.3121 (44.4258)	505.1195 (45.4641)	528.7073 (46.5135)	553.0873 (47.5739)	578.2712 (48.6449)	604.2704 (49.7264)
$n=4$	980.1027 (63.3297)	1011.6628 (64.3413)	1043.9725 (65.3606)	1077.0401 (66.3877)	1110.8741 (67.4224)	1145.4827 (68.4646)	1180.8739 (69.5142)	1217.0558 (70.5711)

Odd Modes $\left.\dfrac{d}{d\xi}Se_m\left(\xi,\bar{q}_{m,n}\right)\right\|_{\xi=\xi_0}=0$	$m=1$	$m=2$	$m=3$	$m=4$	$m=5$	$m=6$	$m=7$	$m=8$
$n=1$	29.2846 (10.9469)	35.1364 (11.9908)	41.7876 (13.0766)	49.2805 (14.2007)	57.6524 (15.3596)	66.9366 (16.5502)	77.1622 (17.7694)	88.3549 (19.0146)
$n=2$	248.8276 (31.9096)	264.9723 (32.9285)	281.8785 (33.9627)	299.5626 (35.0119)	318.0404 (36.0755)	337.3273 (37.1533)	357.4383 (38.2448)	378.3877 (39.3496)
$n=3$	682.7474 (52.8569)	709.1694 (53.8700)	736.3435 (54.8924)	764.2796 (55.9239)	792.9877 (56.9646)	822.4776 (58.0141)	852.7588 (59.0724)	883.8406 (60.1393)
$n=4$	1331.0553 (73.8023)	1367.7533 (74.8127)	1405.1991 (75.8299)	1443.3999 (76.8537)	1482.3630 (77.8841)	1522.0955 (78.9210)	1562.6044 (79.9643)	1603.8965 (81.0139)

Table 2.5 The q parameters and cut-on frequencies for aspect-ratio $D_2/D_1 = 0.2$ ($e = 0.9798$, $\xi_0 = 0.2027$)

Even Modes $\left.\dfrac{d}{d\xi}Ce_m\left(\xi,q_{m,n}\right)\right\|_{\xi=\xi_0}=0$	$m=0$	$m=1$	$m=2$	$m=3$	$m=4$	$m=5$	$m=6$	$m=7$
$n=1$	0 (0)	0.8523 (1.8844)	2.8999 (3.4761)	6.1170 (5.0485)	10.4983 (6.6138)	16.0411 (8.1754)	22.7431 (9.7346)	30.6029 (11.2921)
$n=2$	62.9405 (16.1942)	71.2465 (17.2296)	80.3241 (18.2944)	90.2032 (19.3868)	100.9115 (20.5052)	112.4750 (21.6482)	124.9176 (22.8142)	138.2612 (24.0018)
$n=3$	244.3960 (31.9111)	260.2999 (32.9330)	276.9546 (33.9702)	294.3760 (35.0224)	312.5796 (36.0890)	331.5805 (37.1697)	351.3932 (38.2641)	372.0318 (39.3717)
$n=4$	544.2818 (47.6219)	567.7786 (48.6389)	592.0183 (49.6663)	617.0119 (50.7039)	642.7701 (51.7514)	669.3035 (52.8088)	696.6222 (53.8757)	724.7362 (54.9521)

Odd Modes $\left.\dfrac{d}{d\xi}Se_m\left(\xi,\bar{q}_{m,n}\right)\right\|_{\xi=\xi_0}=0$	$m=1$	$m=2$	$m=3$	$m=4$	$m=5$	$m=6$	$m=7$	$m=8$
$n=1$	16.6120 (8.3197)	21.0984 (9.3760)	26.3846 (10.4850)	32.5199 (11.6404)	39.5464 (12.8365)	47.4992 (14.0682)	56.4079 (15.3308)	66.2972 (16.6204)
$n=2$	138.8651 (24.0542)	150.9715 (25.0808)	163.8360 (26.1276)	177.4794 (27.1937)	191.9218 (28.2785)	207.1822 (29.3813)	223.2788 (30.5013)	240.2289 (31.6379)
$n=3$	379.5347 (39.7668)	399.2353 (40.7858)	419.6819 (41.8172)	440.8877 (42.8606)	462.8653 (43.9159)	485.6270 (44.9827)	509.1848 (46.0609)	533.5506 (47.1501)
$n=4$	738.6375 (55.4766)	765.9302 (56.4923)	793.9636 (57.5168)	822.7472 (58.5501)	852.2901 (59.5920)	882.6016 (60.6424)	913.6904 (61.7012)	945.5655 (62.7683)

Table 2.6 The q parameters and cut-on frequencies for aspect-ratio $D_2/D_1 = 0.25$ ($e = 0.9683$, $\xi_0 = 0.2554$)

Even Modes $\left.\dfrac{d}{d\xi}Ce_m\left(\xi,q_{m,n}\right)\right|_{\xi=\xi_0}=0$	$m=0$	$m=1$	$m=2$	$m=3$	$m=4$	$m=5$	$m=6$	$m=7$
$n=1$	0 (0)	0.8312 (1.8832)	2.8223 (3.4701)	5.9450 (5.0364)	10.1928 (6.5946)	15.5613 (8.1483)	22.0473 (9.6989)	29.6481 (11.2472)
$n=2$	39.9138 (13.0499)	46.5658 (14.0954)	53.9839 (15.1767)	62.2031 (16.2911)	71.2553 (17.4362)	81.1694 (18.6097)	91.9715 (19.8094)	103.6851 (21.0331)
$n=3$	153.9326 (25.6277)	166.5474 (26.6571)	179.9044 (27.7055)	194.0227 (28.7720)	208.9206 (29.8562)	224.6158 (30.9574)	241.1249 (32.0749)	258.4640 (33.2081)
$n=4$	341.9672 (38.1976)	360.5379 (39.2211)	379.8416 (40.2574)	399.8915 (41.3062)	420.7003 (42.3673)	442.2805 (43.4403)	464.6441 (44.5251)	487.8029 (45.6212)
Odd Modes $\left.\dfrac{d}{d\xi}Se_m\left(\xi,\bar{q}_{m,n}\right)\right|_{\xi=\xi_0}=0$	$m=1$	$m=2$	$m=3$	$m=4$	$m=5$	$m=6$	$m=7$	$m=8$
$n=1$	10.6437 (6.7389)	14.2821 (7.8062)	18.7114 (8.9351)	23.9855 (10.1162)	30.1478 (11.3415)	37.2332 (12.6040)	45.2699 (13.8979)	54.2804 (15.2183)
$n=2$	87.6721 (19.3408)	97.3075 (20.3759)	107.6936 (21.4358)	118.8552 (22.5192)	130.8158 (23.6251)	143.5972 (24.7524)	157.2202 (25.8999)	171.7039 (27.0666)
$n=3$	238.6974 (31.9130)	254.2905 (32.9389)	270.6202 (33.9801)	287.7023 (35.0361)	305.5518 (36.1066)	324.1833 (37.1912)	343.6108 (38.2893)	363.8478 (39.4007)
$n=4$	463.7422 (44.4818)	485.2903 (45.5035)	507.5687 (46.5363)	530.5887 (47.5799)	554.3616 (48.6341)	578.8980 (49.6987)	604.2085 (50.7736)	630.3036 (51.8584)

Table 2.7 The q parameters and cut-on frequencies for aspect-ratio $D_2/D_1 = 0.3$ ($e = 0.9539$, $\xi_0 = 0.3095$)

Even Modes $\left.\dfrac{d}{d\xi}Ce_m\left(\xi,q_{m,n}\right)\right|_{\xi=\xi_0}=0$	$m=0$	$m=1$	$m=2$	$m=3$	$m=4$	$m=5$	$m=6$	$m=7$
$n=1$	0 (0)	0.8056 (1.8818)	2.7280 (3.4628)	5.7366 (5.0215)	9.8225 (6.5708)	14.9798 (8.1145)	21.2033 (9.6541)	28.4886 (11.1904)
$n=2$	27.2932 (10.9531)	32.8112 (12.0094)	39.0858 (13.1075)	46.1558 (14.2437)	54.0564 (15.4146)	62.8185 (16.6170)	72.4698 (17.8479)	83.0349 (19.1047)
$n=3$	104.5679 (21.4392)	114.9366 (22.4770)	126.0353 (23.5372)	137.8858 (24.6189)	150.5085 (25.7211)	163.9231 (26.8429)	178.1479 (27.9833)	193.2002 (29.1416)
$n=4$	231.7315 (31.9155)	246.9422 (32.9463)	262.8725 (33.9924)	279.5375 (35.0533)	296.9516 (36.1287)	315.1289 (37.2180)	334.0828 (38.3210)	353.8264 (39.4371)
Odd Modes $\left.\dfrac{d}{d\xi}Se_m\left(\xi,\bar{q}_{m,n}\right)\right|_{\xi=\xi_0}=0$	$m=1$	$m=2$	$m=3$	$m=4$	$m=5$	$m=6$	$m=7$	$m=8$
$n=1$	7.3421 (5.6809)	10.3886 (6.7575)	14.2092 (7.9030)	18.8602 (9.1050)	24.3848 (10.3531)	30.8165 (11.6386)	38.1805 (12.9548)	46.4959 (14.2961)
$n=2$	59.6954 (16.1987)	67.6414 (17.2431)	76.3268 (18.3167)	85.7799 (19.4179)	96.0264 (20.5449)	107.0906 (21.6963)	118.9945 (22.8703)	131.7586 (24.0657)
$n=3$	161.9130 (26.6778)	174.7030 (27.7115)	188.2169 (28.7633)	202.4724 (29.8327)	217.4866 (30.9190)	233.2758 (32.0217)	249.8555 (33.1401)	267.2407 (34.2737)
$n=4$	314.0237 (37.1527)	331.6548 (38.1815)	350.0023 (39.2234)	369.0794 (40.2781)	388.8986 (41.3454)	409.4723 (42.4250)	430.8123 (43.5164)	452.9301 (44.6195)

Table 2.8 The q parameters and cut-on frequencies for aspect-ratio $D_2/D_1 = 0.35$ ($e = 0.9368$, $\xi_0 = 0.3654$)

Even Modes $\dfrac{\mathrm{d}}{\mathrm{d}\xi} Ce_m\left(\xi, q_{m,n}\right)\bigg|_{\xi=\xi_0} = 0$	$m=0$	$m=1$	$m=2$	$m=3$	$m=4$	$m=5$	$m=6$	$m=7$
$n=1$	0 (0)	0.7754 (1.8800)	2.6174 (3.4542)	5.4924 (5.0037)	9.3890 (6.5421)	14.2984 (8.0733)	20.2129 (9.5989)	27.1246 (11.1196)
$n=2$	19.6119 (9.4551)	24.2921 (10.5230)	29.7151 (11.6384)	35.9236 (12.7967)	42.9550 (13.9931)	50.8425 (15.2237)	59.6153 (16.4849)	69.2998 (17.7735)
$n=3$	74.6596 (18.4480)	83.3763 (19.4952)	92.8070 (20.5682)	102.9755 (21.6657)	113.9041 (22.7864)	125.6137 (23.9290)	138.1236 (25.0923)	151.4519 (26.2751)
$n=4$	165.0480 (27.4291)	177.7912 (28.4683)	191.2370 (29.5252)	205.4019 (30.5991)	220.3019 (31.6895)	235.9522 (32.7958)	252.3673 (33.9174)	269.5611 (35.0538)
Odd Modes $\dfrac{\mathrm{d}}{\mathrm{d}\xi} Se_m\left(\xi, \bar{q}_{m,n}\right)\bigg|_{\xi=\xi_0} = 0$	$m=1$	$m=2$	$m=3$	$m=4$	$m=5$	$m=6$	$m=7$	$m=8$
$n=1$	5.3130 (4.9213)	7.9127 (6.0058)	11.2635 (7.1654)	15.4223 (8.3846)	20.4304 (9.6504)	26.3176 (10.9529)	33.1044 (12.2843)	40.8044 (13.6383)
$n=2$	42.7193 (13.9546)	49.4208 (15.0093)	56.8466 (16.0975)	65.0273 (17.2169)	73.9908 (18.3652)	83.7624 (19.5403)	94.3652 (20.7402)	105.8199 (21.9629)
$n=3$	115.4355 (22.9391)	126.1659 (23.9816)	137.6034 (25.0450)	149.7678 (26.1286)	162.6776 (27.2314)	176.3503 (28.3527)	190.8028 (29.4916)	206.0505 (30.6474)
$n=4$	223.4975 (31.9185)	238.2531 (32.9553)	253.7079 (34.0074)	269.8763 (35.0743)	286.7721 (36.1555)	304.4088 (37.2508)	322.7992 (38.3595)	341.9556 (39.4813)

Table 2.9 The q parameters and cut-on frequencies for aspect-ratio $D_2/D_1 = 0.4$ ($e = 0.9165$, $\xi_0 = 0.4237$)

Even Modes $\dfrac{\mathrm{d}}{\mathrm{d}\xi} Ce_m\left(\xi, q_{m,n}\right)\bigg|_{\xi=\xi_0} = 0$	$m=0$	$m=1$	$m=2$	$m=3$	$m=4$	$m=5$	$m=6$	$m=7$
$n=1$	0 (0)	0.7407 (1.8781)	2.4909 (3.4441)	5.2135 (4.9826)	8.8936 (6.5077)	13.5187 (8.0234)	19.0766 (9.5310)	25.5543 (11.0312)
$n=2$	14.5775 (8.3317)	18.6040 (9.4122)	23.3562 (10.5461)	28.8803 (11.7271)	35.2165 (12.9498)	42.4005 (14.2094)	50.4640 (15.5018)	59.4353 (16.8234)
$n=3$	55.1489 (16.2054)	62.5830 (17.2631)	70.7112 (18.3499)	79.5587 (19.4641)	89.1491 (20.6039)	99.5040 (21.7676)	110.6433 (22.9537)	122.5851 (24.1607)
$n=4$	121.6188 (24.0653)	132.4498 (25.1140)	143.9626 (26.1827)	156.1750 (27.2707)	169.1040 (28.3771)	182.7654 (29.5010)	197.1746 (30.6419)	212.3457 (31.7989)
Odd Modes $\dfrac{\mathrm{d}}{\mathrm{d}\xi} Se_m\left(\xi, \bar{q}_{m,n}\right)\bigg|_{\xi=\xi_0} = 0$	$m=1$	$m=2$	$m=3$	$m=4$	$m=5$	$m=6$	$m=7$	$m=8$
$n=1$	3.9695 (4.3477)	6.2132 (5.4394)	9.1796 (6.6115)	12.9240 (7.8449)	17.4834 (9.1244)	22.8813 (10.4383)	29.1316 (11.7780)	36.2419 (13.1370)
$n=2$	31.6272 (12.2722)	37.3606 (13.3382)	43.7992 (14.4419)	50.9753 (15.5801)	58.9181 (16.7500)	67.6536 (17.9488)	77.2048 (19.1740)	87.5922 (20.4232)
$n=3$	85.1458 (20.1359)	94.2788 (21.1884)	104.0985 (22.2645)	114.6261 (23.3632)	125.8810 (24.4833)	137.8817 (25.6238)	150.6455 (26.7836)	164.1884 (27.9616)
$n=4$	164.5683 (27.9939)	177.0969 (29.0399)	190.3037 (30.1033)	204.2041 (31.1833)	218.8129 (32.2795)	234.1444 (33.3912)	250.2119 (34.5179)	267.0283 (35.6590)

Table 2.10   The q parameters and cut-on frequencies for aspect-ratio $D_2/D_1 = 0.45$ ($e = 0.8930$, $\xi_0 = 0.4847$)

Even Modes $\left.\dfrac{\mathrm{d}}{\mathrm{d}\xi}Ce_m\left(\xi, q_{m,n}\right)\right|_{\xi=\xi_0}=0$	$m=0$	$m=1$	$m=2$	$m=3$	$m=4$	$m=5$	$m=6$	$m=7$
$n=1$	0 (0)	0.7016 (1.8759)	2.3489 (3.4324)	4.9008 (4.9579)	8.3379 (6.4669)	12.6420 (7.9629)	17.7946 (9.4473)	23.7760 (10.9203)
$n=2$	11.0903 (7.4582)	14.5852 (8.5530)	18.7863 (9.7070)	23.7431 (10.9127)	29.4994 (12.1639)	36.0940 (13.4550)	43.5607 (14.7813)	51.9281 (16.1386)
$n=3$	41.6995 (14.4621)	48.0952 (15.5316)	55.1611 (16.6334)	62.9233 (17.7652)	71.4061 (18.9248)	80.6313 (20.1102)	90.6188 (21.3193)	101.3864 (22.5504)
$n=4$	91.7337 (21.4501)	101.0202 (22.5097)	110.9639 (23.5915)	121.5834 (24.6946)	132.8963 (25.8179)	144.9188 (26.9604)	157.6665 (28.1212)	171.1537 (29.2993)
Odd Modes $\left.\dfrac{\mathrm{d}}{\mathrm{d}\xi}Se_m\left(\xi, \bar{q}_{m,n}\right)\right|_{\xi=\xi_0}=0$	$m=1$	$m=2$	$m=3$	$m=4$	$m=5$	$m=6$	$m=7$	$m=8$
$n=1$	3.0296 (3.8981)	4.9778 (4.9967)	7.6153 (6.1803)	10.9926 (7.4253)	15.1397 (8.7141)	20.0721 (10.0337)	25.7965 (11.3748)	32.3142 (12.7310)
$n=2$	23.9681 (10.9643)	28.9162 (12.0430)	34.5462 (13.1633)	40.8917 (14.3213)	47.9822 (15.5133)	55.8435 (16.7360)	64.4981 (17.9861)	73.9651 (19.2610)
$n=3$	64.2877 (17.9568)	72.1292 (19.0204)	80.6332 (20.1104)	89.8215 (21.2253)	99.7144 (22.3637)	110.3306 (23.5241)	121.6875 (24.7052)	133.8011 (25.9056)
$n=4$	124.0379 (24.9426)	134.7689 (25.9992)	146.1533 (27.0750)	158.2073 (28.1694)	170.9462 (29.2816)	184.3846 (30.4107)	198.5364 (31.5562)	213.4144 (32.7172)

Table 2.11   The q parameters and cut-on frequencies for aspect-ratio $D_2/D_1 = 0.5$ ($e = 0.8660$, $\xi_0 = 0.5493$)

Even Modes $\left.\dfrac{\mathrm{d}}{\mathrm{d}\xi}Ce_m\left(\xi, q_{m,n}\right)\right|_{\xi=\xi_0}=0$	$m=0$	$m=1$	$m=2$	$m=3$	$m=4$	$m=5$	$m=6$	$m=7$
$n=1$	0 (0)	0.6582 (1.8736)	2.1918 (3.4190)	4.5554 (4.9291)	7.7233 (6.4180)	11.6701 (7.8893)	16.3693 (9.3436)	21.7950 (10.7815)
$n=2$	8.5689 (6.7602)	11.6175 (7.8715)	15.3504 (9.0481)	19.8214 (10.2817)	25.0777 (11.5649)	31.1593 (12.8912)	38.0981 (14.2545)	45.9169 (15.6490)
$n=3$	32.0226 (13.0686)	37.5490 (14.1514)	43.7174 (15.2696)	50.5545 (16.4202)	58.0846 (17.6007)	66.3293 (18.8084)	75.3081 (20.0410)	85.0388 (21.2965)
$n=4$	70.2713 (19.3592)	78.2681 (20.4311)	86.8936 (21.5275)	96.1665 (22.6470)	106.1048 (23.7885)	116.7247 (24.9506)	128.0417 (26.1321)	140.0699 (27.3320)
Odd Modes $\left.\dfrac{\mathrm{d}}{\mathrm{d}\xi}Se_m\left(\xi, \bar{q}_{m,n}\right)\right|_{\xi=\xi_0}=0$	$m=1$	$m=2$	$m=3$	$m=4$	$m=5$	$m=6$	$m=7$	$m=8$
$n=1$	2.3436 (3.5354)	4.0390 (4.6413)	6.3842 (5.8352)	9.4225 (7.0890)	13.1745 (8.3824)	17.6473 (9.7015)	22.8408 (11.0371)	28.7504 (12.3829)
$n=2$	18.4477 (9.9191)	22.7368 (11.0119)	27.6803 (12.1502)	33.3121 (13.3291)	39.6613 (14.5440)	46.7522 (15.7907)	54.6054 (17.0654)	63.2370 (18.3647)
$n=3$	49.2969 (16.2147)	56.0588 (17.2911)	63.4549 (18.3964)	71.5071 (19.5287)	80.2361 (20.6864)	89.6605 (21.8675)	99.7974 (23.0706)	110.6626 (24.2940)
$n=4$	94.9464 (22.5029)	104.1777 (23.5715)	114.0336 (24.6613)	124.5306 (25.7714)	135.6844 (26.9007)	147.5096 (28.0485)	160.0200 (29.2137)	173.2286 (30.3955)

Table 2.12 The q parameters and cut-on frequencies for aspect-ratio $D_2/D_1 = 0.55$ ($e = 0.8352$, $\xi_0 = 0.6184$)

Even Modes $\left.\dfrac{d}{d\xi}Ce_m\left(\xi,q_{m,n}\right)\right|_{\xi=\xi_0}=0$	$m=0$	$m=1$	$m=2$	$m=3$	$m=4$	$m=5$	$m=6$	$m=7$
$n=1$	0 (0)	0.6104 (1.8710)	2.0202 (3.4037)	4.1785 (4.8952)	7.0519 (6.3593)	10.6068 (7.7992)	14.8103 (9.2159)	19.6348 (10.6113)
$n=2$	6.6823 (6.1904)	9.3461 (7.3211)	12.6705 (8.5242)	16.7129 (9.7900)	21.5212 (11.1094)	27.1306 (12.4735)	33.5627 (13.8735)	40.8254 (15.3011)
$n=3$	24.8172 (11.9298)	29.5950 (13.0277)	34.9823 (14.1639)	41.0058 (15.3349)	47.6894 (16.5375)	55.0544 (17.7686)	63.1202 (19.0258)	71.9049 (20.3066)
$n=4$	54.3222 (17.6501)	61.2119 (18.7359)	68.6972 (19.8485)	76.7973 (20.9861)	85.5296 (22.1471)	94.9103 (23.3300)	104.9543 (24.5334)	115.6755 (25.7560)
Odd Modes $\left.\dfrac{d}{d\xi}Se_m\left(\xi,\bar{q}_{m,n}\right)\right|_{\xi=\xi_0}=0$	$m=1$	$m=2$	$m=3$	$m=4$	$m=5$	$m=6$	$m=7$	$m=8$
$n=1$	1.8259 (3.2359)	3.2992 (4.3497)	5.3766 (5.5528)	8.0914 (6.8119)	11.4542 (8.1048)	15.4643 (9.4172)	20.1155 (10.7405)	25.3994 (12.0689)
$n=2$	14.3295 (9.0651)	18.0497 (10.1740)	22.3922 (11.3320)	27.3900 (12.5330)	33.0703 (13.7714)	39.4544 (15.0420)	46.5584 (16.3402)	54.3932 (17.6616)
$n=3$	38.1478 (14.7908)	43.9818 (15.8816)	50.4172 (17.0038)	57.4760 (18.1552)	65.1785 (19.3335)	73.5429 (20.5366)	82.5859 (21.7626)	92.3226 (23.0098)
$n=4$	73.3408 (20.5083)	81.2857 (21.5906)	89.8223 (22.6960)	98.9672 (23.8234)	108.7357 (24.9715)	119.1425 (26.1391)	130.2010 (27.3253)	141.9237 (28.5289)

Table 2.13 The q parameters and cut-on frequencies for aspect-ratio $D_2/D_1 = 0.6$ ($e = 0.8000$, $\xi_0 = 0.6931$)

Even Modes $\left.\dfrac{d}{d\xi}Ce_m\left(\xi,q_{m,n}\right)\right|_{\xi=\xi_0}=0$	$m=0$	$m=1$	$m=2$	$m=3$	$m=4$	$m=5$	$m=6$	$m=7$
$n=1$	0 (0)	0.5584 (1.8682)	1.8346 (3.3862)	3.7714 (4.8550)	6.3271 (6.2885)	9.4614 (7.6899)	13.1410 (9.0626)	17.3450 (10.4118)
$n=2$	5.2304 (5.7175)	7.5548 (6.8715)	10.5134 (8.1061)	14.1640 (9.4088)	18.5478 (10.7668)	23.6860 (12.1671)	29.5825 (13.5974)	36.2289 (15.0476)
$n=3$	19.2987 (10.9826)	23.4162 (12.0976)	28.1061 (13.2538)	33.3947 (14.4470)	39.3053 (15.6735)	45.8598 (16.9300)	53.0786 (18.2138)	60.9831 (19.5229)
$n=4$	42.1332 (16.2275)	48.0496 (17.3295)	54.5243 (18.4602)	61.5760 (19.6176)	69.2217 (20.7999)	77.4770 (22.0053)	86.3563 (23.2320)	95.8729 (24.4787)
Odd Modes $\left.\dfrac{d}{d\xi}Se_m\left(\xi,\bar{q}_{m,n}\right)\right|_{\xi=\xi_0}=0$	$m=1$	$m=2$	$m=3$	$m=4$	$m=5$	$m=6$	$m=7$	$m=8$
$n=1$	1.4247 (2.9841)	2.6983 (4.1066)	4.5237 (5.3173)	6.9227 (6.5777)	9.8964 (7.8646)	13.4375 (9.1643)	17.5358 (10.4690)	22.1811 (11.7742)
$n=2$	11.1693 (8.3551)	14.3860 (9.4822)	18.1877 (10.6618)	22.6046 (11.8861)	27.6599 (13.1482)	33.3694 (14.4416)	39.7422 (15.7604)	46.7817 (17.0993)
$n=3$	29.6197 (13.6060)	34.6369 (14.7133)	40.2180 (15.8544)	46.3848 (17.0266)	53.1569 (18.2272)	60.5517 (19.4538)	68.5853 (20.7041)	77.2719 (21.9761)
$n=4$	56.8396 (18.8480)	63.6549 (19.9460)	71.0246 (21.0690)	78.9647 (22.2155)	87.4905 (23.3841)	96.6160 (24.5734)	106.3541 (25.7820)	116.7170 (27.0089)

Table 2.14   The q parameters and cut-on frequencies for aspect-ratio $D_2/D_1 = 0.65$ ($e = 0.7599$, $\xi_0 = 0.7753$)

Even Modes $\dfrac{\mathrm{d}}{\mathrm{d}\xi} Ce_m\left(\xi, q_{m,n}\right)\Big|_{\xi=\xi_0} = 0$	$m = 0$	$m = 1$	$m = 2$	$m = 3$	$m = 4$	$m = 5$	$m = 6$	$m = 7$
$n = 1$	0 (0)	0.5023 (1.8653)	1.6356 (3.3658)	3.3364 (4.8072)	5.5555 (6.2032)	8.2520 (7.5602)	11.4009 (8.8864)	14.9928 (10.1905)
$n = 2$	4.0864 (5.3202)	6.1050 (6.5027)	8.7257 (7.7742)	11.9996 (9.1167)	15.9517 (10.5113)	20.5833 (11.9402)	25.8811 (13.3889)	31.8229 (14.8465)
$n = 3$	14.9710 (10.1831)	18.4935 (11.3178)	22.5465 (12.4966)	27.1556 (13.7146)	32.3451 (14.9678)	38.1388 (16.2532)	44.5628 (17.5687)	51.6459 (18.9135)
$n = 4$	32.5968 (15.0259)	37.6394 (16.1464)	43.1983 (17.2977)	49.2911 (18.4773)	55.9344 (19.6831)	63.1429 (20.9130)	70.9308 (22.1652)	79.3113 (23.4381)
Odd Modes $\dfrac{\mathrm{d}}{\mathrm{d}\xi} Se_m\left(\xi, \bar{q}_{m,n}\right)\Big|_{\xi=\xi_0} = 0$	$m = 1$	$m = 2$	$m = 3$	$m = 4$	$m = 5$	$m = 6$	$m = 7$	$m = 8$
$n = 1$	1.1070 (2.7690)	2.1973 (3.9012)	3.7807 (5.1173)	5.8670 (6.3747)	8.4495 (7.6501)	11.5166 (8.9313)	15.0565 (10.2121)	19.0595 (11.4897)
$n = 2$	8.6857 (7.7563)	11.4469 (8.9043)	14.7489 (10.1073)	18.6174 (11.3557)	23.0686 (12.6405)	28.1103 (13.9536)	33.7427 (15.2878)	39.9599 (16.6367)
$n = 3$	22.9410 (12.6055)	27.2237 (13.7318)	32.0281 (14.8943)	37.3751 (16.0896)	43.2832 (17.3146)	49.7689 (18.5666)	56.8468 (19.8430)	64.5294 (21.1413)
$n = 4$	43.9384 (17.4452)	49.7410 (18.5614)	56.0557 (19.7044)	62.8981 (20.8724)	70.2828 (22.0637)	78.2231 (23.2767)	86.7314 (24.5099)	95.8194 (25.7621)

Table 2.15   The q parameters and cut-on frequencies for aspect-ratio $D_2/D_1 = 0.7$ ($e = 0.7141$, $\xi_0 = 0.8673$)

Even Modes $\dfrac{\mathrm{d}}{\mathrm{d}\xi} Ce_m\left(\xi, q_{m,n}\right)\Big|_{\xi=\xi_0} = 0$	$m = 0$	$m = 1$	$m = 2$	$m = 3$	$m = 4$	$m = 5$	$m = 6$	$m = 7$
$n = 1$	0 (0)	0.4421 (1.8622)	1.4240 (3.3419)	2.8765 (4.7499)	4.7479 (6.1024)	7.0053 (7.4124)	9.6366 (8.6938)	12.6405 (9.9570)
$n = 2$	3.1665 (4.9835)	4.9027 (6.2010)	7.1985 (7.5139)	10.0885 (8.8953)	13.5758 (10.3188)	17.6438 (11.7636)	22.2649 (13.2146)	27.4023 (14.6601)
$n = 3$	11.5076 (9.5003)	14.4835 (10.6581)	17.9429 (11.8629)	21.9122 (13.1096)	26.4187 (14.3946)	31.4931 (15.7164)	37.1689 (17.0740)	43.4751 (18.4657)
$n = 4$	24.9847 (13.9985)	29.2279 (15.1406)	33.9399 (16.3155)	39.1376 (17.5203)	44.8365 (18.7526)	51.5013 (20.0101)	57.7965 (21.2910)	65.0877 (22.5941)
Odd Modes $\dfrac{\mathrm{d}}{\mathrm{d}\xi} Se_m\left(\xi, \bar{q}_{m,n}\right)\Big|_{\xi=\xi_0} = 0$	$m = 1$	$m = 2$	$m = 3$	$m = 4$	$m = 5$	$m = 6$	$m = 7$	$m = 8$
$n = 1$	0.8507 (2.5831)	1.7698 (3.7257)	3.1171 (4.9445)	4.8919 (6.1942)	7.0820 (7.4529)	9.6742 (8.7107)	12.6578 (9.9638)	16.0253 (11.2111)
$n = 2$	6.6935 (7.2456)	9.0334 (8.4172)	11.8620 (9.6455)	15.1974 (10.9177)	19.0468 (12.2224)	23.4077 (13.5495)	28.2711 (14.8907)	33.6232 (16.2392)
$n = 3$	17.6044 (11.7505)	21.2137 (12.8989)	25.2973 (14.0858)	29.8744 (15.3072)	34.9621 (16.5594)	40.5748 (17.8391)	46.7237 (19.1432)	53.4159 (20.4682)
$n = 4$	33.6487 (16.2454)	38.5258 (17.3828)	43.8678 (18.5489)	49.6894 (19.7413)	56.0041 (20.9582)	62.8247 (22.1978)	70.1631 (23.4584)	78.0303 (24.7387)

Table 2.16 The q parameters and cut-on frequencies for aspect-ratio $D_2/D_1 = 0.75$ ($e = 0.6614$, $\xi_0 = 0.9730$)

Even Modes $\dfrac{d}{d\xi} Ce_m\left(\xi, q_{m,n}\right)\Big\|_{\xi=\xi_0} = 0$	$m=0$	$m=1$	$m=2$	$m=3$	$m=4$	$m=5$	$m=6$	$m=7$
$n=1$	0 (0)	0.3780 (1.8590)	1.2009 (3.3136)	2.3971 (4.6815)	3.9201 (5.9868)	5.7513 (7.2515)	7.8883 (8.4924)	10.3317 (9.7191)
$n=2$	2.4131 (4.6971)	3.8808 (5.9566)	5.8489 (7.3127)	8.3288 (8.7263)	11.3042 (10.1662)	14.7440 (11.6104)	18.6060 (13.0427)	22.8422 (14.4514)
$n=3$	8.6865 (8.9118)	11.1509 (10.0971)	14.0475 (11.3329)	17.4059 (12.6151)	21.2601 (13.9420)	25.6449 (15.3123)	30.5806 (16.7211)	36.0603 (18.1575)
$n=4$	18.8020 (13.1112)	22.3005 (14.2790)	26.2147 (15.4815)	30.5605 (16.7156)	35.3533 (17.9786)	40.6094 (19.2688)	46.3483 (20.5853)	52.5951 (21.9287)
Odd Modes $\dfrac{d}{d\xi} Se_m\left(\xi, \bar{q}_{m,n}\right)\Big\|_{\xi=\xi_0} = 0$	$m=1$	$m=2$	$m=3$	$m=4$	$m=5$	$m=6$	$m=7$	$m=8$
$n=1$	0.6409 (2.4207)	1.3973 (3.5743)	2.5122 (4.7926)	3.9765 (6.0297)	5.7757 (7.2668)	7.8980 (8.4976)	10.3353 (9.7208)	13.0828 (10.9368)
$n=2$	5.0663 (6.8059)	7.0081 (8.0047)	9.3769 (9.2591)	12.1805 (10.5530)	15.4157 (11.8720)	19.0707 (13.2046)	23.1290 (14.5418)	27.5727 (15.8774)
$n=3$	13.2646 (11.0126)	16.2454 (12.1873)	19.6467 (13.4025)	23.4858 (14.6536)	27.7762 (15.9359)	32.5274 (17.2451)	37.7422 (18.5761)	43.4170 (19.9237)
$n=4$	25.2987 (15.2086)	29.3150 (16.3714)	33.7432 (17.5644)	38.5969 (18.7853)	43.8889 (20.0317)	49.6308 (21.3018)	55.8336 (22.5938)	62.5068 (23.9059)

Table 2.17 The q parameters and cut-on frequencies for aspect-ratio $D_2/D_1 = 0.8$ ($e = 0.6000$, $\xi_0 = 1.0986$)

Even Modes $\dfrac{d}{d\xi} Ce_m\left(\xi, q_{m,n}\right)\Big\|_{\xi=\xi_0} = 0$	$m=0$	$m=1$	$m=2$	$m=3$	$m=4$	$m=5$	$m=6$	$m=7$
$n=1$	0 (0)	0.3099 (1.8556)	0.9679 (3.2795)	1.9055 (4.6014)	3.0901 (5.8596)	4.5157 (7.0834)	6.1829 (8.2885)	8.0916 (9.4819)
$n=2$	1.7852 (4.4537)	2.9880 (5.7619)	4.6113 (7.1580)	6.6447 (8.5925)	9.0576 (10.0319)	11.8055 (11.4531)	14.8408 (12.8412)	18.1334 (14.1944)
$n=3$	6.3520 (8.4010)	8.3296 (9.6203)	10.6847 (10.8958)	13.4523 (12.2258)	16.6652 (13.6077)	20.3318 (15.0303)	24.4286 (16.4751)	28.9117 (17.9232)
$n=4$	13.7024 (12.3389)	16.4948 (13.5379)	19.6437 (14.7737)	23.1648 (16.0433)	27.0764 (17.3450)	31.4029 (18.6794)	36.1759 (20.0488)	41.4232 (21.4536)
Odd Modes $\dfrac{d}{d\xi} Se_m\left(\xi, \bar{q}_{m,n}\right)\Big\|_{\xi=\xi_0} = 0$	$m=1$	$m=2$	$m=3$	$m=4$	$m=5$	$m=6$	$m=7$	$m=8$
$n=1$	0.4668 (2.2775)	1.0666 (3.4426)	1.9516 (4.6566)	3.1077 (5.8762)	4.5216 (7.0881)	6.1848 (8.2897)	8.0921 (9.4822)	10.2408 (10.6671)
$n=2$	3.7152 (6.4249)	5.2727 (7.6541)	7.1842 (8.9345)	9.4467 (10.2452)	12.0475 (11.5698)	14.9687 (12.8965)	18.1920 (14.2174)	21.7020 (15.5285)
$n=3$	9.6801 (10.3710)	12.0633 (11.5774)	14.8058 (12.8261)	17.9206 (14.1109)	21.4141 (15.4251)	25.2851 (16.7614)	29.5236 (18.1119)	34.1133 (19.4689)
$n=4$	18.4183 (14.3055)	21.6199 (15.4991)	25.1739 (16.7245)	29.0927 (17.9792)	33.3876 (19.2607)	38.0681 (20.5665)	43.1412 (21.8940)	48.6084 (23.2399)

Table 2.18 The q parameters and cut-on frequencies for aspect-ratio $D_2/D_1 = 0.85$ ($e = 0.5268$, $\xi_0 = 1.2562$)

Even Modes $\left.\dfrac{d}{d\xi} Ce_m\left(\xi, q_{m,n}\right)\right|_{\xi=\xi_0} = 0$	$m = 0$	$m = 1$	$m = 2$	$m = 3$	$m = 4$	$m = 5$	$m = 6$	$m = 7$
$n = 1$	0	0.2380	0.7273	1.4113	2.2741	3.3156	4.5357	5.9334
	(0)	(1.8521)	(3.2379)	(4.5103)	(5.7254)	(6.9133)	(8.0858)	(9.2480)
$n = 2$	1.2523	2.1830	3.4355	4.9842	6.7878	8.8019	11.0048	13.3969
	(4.2487)	(5.6095)	(7.0371)	(8.4761)	(9.8915)	(11.2638)	(12.5947)	(13.8963)
$n = 3$	4.3919	5.8983	7.7231	9.8953	12.4139	15.2462	18.3430	21.6529
	(7.9565)	(9.2206)	(10.5510)	(11.9430)	(13.3768)	(14.8245)	(16.2605)	(17.6668)
$n = 4$	9.4368	11.5474	13.9491	16.6614	19.7118	23.1283	26.9121	31.0281
	(11.6630)	(12.9015)	(14.1799)	(15.4972)	(16.8563)	(18.2587)	(19.6957)	(21.1483)
Odd Modes $\left.\dfrac{d}{d\xi} Se_m\left(\xi, \bar{q}_{m,n}\right)\right|_{\xi=\xi_0} = 0$	$m = 1$	$m = 2$	$m = 3$	$m = 4$	$m = 5$	$m = 6$	$m = 7$	$m = 8$
$n = 1$	0.3207	0.7680	1.4252	2.2780	3.3166	4.5359	5.9334	7.5073
	(2.1502)	(3.3272)	(4.5325)	(5.7303)	(6.9142)	(8.0860)	(9.2481)	(10.4026)
$n = 2$	2.5758	3.7541	5.2024	6.9086	8.8550	11.0241	13.4031	15.9838
	(6.0933)	(7.3562)	(8.6597)	(9.9792)	(11.2978)	(12.6058)	(13.8996)	(15.1788)
$n = 3$	6.6770	8.4805	10.5718	12.9551	15.6250	18.5671	21.7625	25.1921
	(9.8104)	(11.0563)	(12.3445)	(13.6653)	(15.0075)	(16.3595)	(17.7114)	(19.0560)
$n = 4$	12.6702	15.0862	17.7876	20.7841	24.0829	27.6854	31.5852	35.7676
	(13.5142)	(14.7465)	(16.0124)	(17.3087)	(18.6317)	(19.9767)	(21.3373)	(22.7061)

Table 2.19 The q parameters and cut-on frequencies for aspect-ratio $D_2/D_1 = 0.9$ ($e = 0.4359$, $\xi_0 = 1.4722$)

Even Modes $\left.\dfrac{d}{d\xi} Ce_m\left(\xi, q_{m,n}\right)\right|_{\xi=\xi_0} = 0$	$m = 0$	$m = 1$	$m = 2$	$m = 3$	$m = 4$	$m = 5$	$m = 6$	$m = 7$
$n = 1$	0	0.1623	0.4825	0.9241	1.4835	2.1605	2.9543	3.8637
	(0)	(1.8486)	(3.1871)	(4.4109)	(5.5885)	(6.7442)	(7.8864)	(9.0189)
$n = 2$	0.7902	1.4327	2.2857	3.3165	4.4874	5.7834	7.2066	8.7589
	(4.0786)	(5.4919)	(6.9368)	(8.3558)	(9.7196)	(11.0343)	(12.3174)	(13.5793)
$n = 3$	2.7231	3.7631	5.0431	6.5573	8.2723	10.1438	12.1406	14.2600
	(7.5716)	(8.9008)	(10.3039)	(11.7494)	(13.1968)	(14.6135)	(15.9873)	(17.3266)
$n = 4$	5.8218	7.2607	8.9209	10.8271	12.9880	15.3765	17.9449	20.6468
	(11.0709)	(12.3635)	(13.7043)	(15.0976)	(16.5357)	(17.9921)	(19.4367)	(20.8487)
Odd Modes $\left.\dfrac{d}{d\xi} Se_m\left(\xi, \bar{q}_{m,n}\right)\right|_{\xi=\xi_0} = 0$	$m = 1$	$m = 2$	$m = 3$	$m = 4$	$m = 5$	$m = 6$	$m = 7$	$m = 8$
$n = 1$	0.1970	0.4941	0.9267	1.4839	2.1606	2.9543	3.8637	4.8879
	(2.0364)	(3.2253)	(4.4170)	(5.5893)	(6.7443)	(7.8864)	(9.0189)	(10.1441)
$n = 2$	1.6000	2.3965	3.3705	4.5060	5.7884	7.2077	8.7591	10.4399
	(5.8039)	(7.1031)	(8.4236)	(9.7398)	(11.0390)	(12.3183)	(13.5795)	(14.8252)
$n = 3$	4.1263	5.3536	6.7818	8.4029	10.2014	12.1601	14.2655	16.5094
	(9.3204)	(10.6164)	(11.9488)	(13.3005)	(14.6549)	(16.0001)	(17.3299)	(18.6432)
$n = 4$	7.8059	9.4477	11.2966	13.3551	15.6182	18.0721	20.6988	23.4816
	(12.8193)	(14.1032)	(15.4215)	(16.7679)	(18.1329)	(19.5055)	(20.8750)	(22.2340)

Table 2.20 The q parameters and cut-on frequencies for aspect-ratio $D_2/D_1 = 0.95$ ($e = 0.3123$, $\xi_0 = 1.8318$)

Even Modes $\left.\dfrac{\mathrm{d}}{\mathrm{d}\xi}Ce_m\left(\xi,q_{m,n}\right)\right|_{\xi=\xi_0}=0$	$m=0$	$m=1$	$m=2$	$m=3$	$m=4$	$m=5$	$m=6$	$m=7$
$n=1$	0 (0)	0.0830 (1.8449)	0.2382 (3.1258)	0.4521 (4.3066)	0.7245 (5.4519)	1.0547 (6.5780)	1.4419 (7.6914)	1.8856 (8.7954)
$n=2$	0.3785 (3.9406)	0.7112 (5.4015)	1.1401 (6.8392)	1.6408 (8.2046)	2.2047 (9.5104)	2.8332 (10.7811)	3.5263 (12.0279)	4.2834 (13.2562)
$n=3$	1.2809 (7.2492)	1.8339 (8.6738)	2.5055 (10.1386)	3.2714 (11.5850)	4.1057 (12.9783)	5.0021 (14.3253)	5.9640 (15.6421)	6.9923 (16.9371)
$n=4$	2.7174 (10.5586)	3.4762 (11.9422)	4.3673 (13.3855)	5.3790 (14.8552)	6.4815 (16.3067)	7.6485 (17.7139)	8.8758 (19.0824)	10.1684 (20.4246)
Odd Modes $\left.\dfrac{\mathrm{d}}{\mathrm{d}\xi}Se_m\left(\xi,\bar{q}_{m,n}\right)\right|_{\xi=\xi_0}=0$	$m=1$	$m=2$	$m=3$	$m=4$	$m=5$	$m=6$	$m=7$	$m=8$
$n=1$	0.0912 (1.9339)	0.2396 (3.1349)	0.4522 (4.3073)	0.7245 (5.4520)	1.0547 (6.5780)	1.4420 (7.6914)	1.8856 (8.7954)	2.3852 (9.8922)
$n=2$	0.7512 (5.5513)	1.1566 (6.8883)	1.6446 (8.2141)	2.2053 (9.5117)	2.8332 (10.7813)	3.5263 (12.0279)	4.2834 (13.2562)	5.1035 (14.4698)
$n=3$	1.9287 (8.8952)	2.5633 (10.2548)	3.2950 (11.6267)	4.1117 (12.9879)	5.0032 (14.3269)	5.9641 (15.6423)	6.9924 (16.9371)	8.0869 (18.2146)
$n=4$	3.6356 (12.2129)	4.4894 (13.5713)	5.4510 (14.9542)	6.5109 (16.3436)	7.6564 (17.7231)	8.8774 (19.0841)	10.1687 (20.4249)	11.5282 (21.7474)

Table 2.21 The q parameters and cut-on frequencies for aspect-ratio $D_2/D_1 = 0.96$ ($e = 0.2800$, $\xi_0 = 1.9459$)

Even Modes $\left.\dfrac{\mathrm{d}}{\mathrm{d}\xi}Ce_m\left(\xi,q_{m,n}\right)\right|_{\xi=\xi_0}=0$	$m=0$	$m=1$	$m=2$	$m=3$	$m=4$	$m=5$	$m=6$	$m=7$
$n=1$	0 (0)	0.0667 (1.8442)	0.1899 (3.1123)	0.3600 (4.2855)	0.5768 (5.4248)	0.8397 (6.5452)	1.1479 (7.6529)	1.5011 (8.7514)
$n=2$	0.3007 (3.9167)	0.5686 (5.3860)	0.9109 (6.8171)	1.3080 (8.1692)	1.7562 (9.4657)	2.2563 (10.7292)	2.8080 (11.9694)	3.4107 (13.1914)
$n=3$	1.0144 (7.1939)	1.4631 (8.6400)	2.0035 (10.1105)	2.6129 (11.5460)	3.2737 (12.9238)	3.9856 (14.2600)	4.7508 (15.5688)	5.5692 (16.8566)
$n=4$	2.1474 (10.4672)	2.7654 (11.8781)	3.4891 (13.3423)	4.3015 (14.8144)	5.1763 (16.2510)	6.0995 (17.6408)	7.0737 (18.9975)	8.1016 (20.3310)
Odd Modes $\left.\dfrac{\mathrm{d}}{\mathrm{d}\xi}Se_m\left(\xi,\bar{q}_{m,n}\right)\right|_{\xi=\xi_0}=0$	$m=1$	$m=2$	$m=3$	$m=4$	$m=5$	$m=6$	$m=7$	$m=8$
$n=1$	0.0718 (1.9146)	0.1906 (3.1181)	0.3600 (4.2859)	0.5768 (5.4248)	0.8397 (6.5452)	1.1479 (7.6529)	1.5011 (8.7514)	1.8988 (9.8427)
$n=2$	0.5940 (5.5049)	0.9195 (6.8494)	1.3096 (8.1741)	1.7563 (9.4662)	2.2563 (10.7293)	2.8080 (11.9694)	3.4107 (13.1914)	4.0636 (14.3988)
$n=3$	1.5240 (8.8179)	2.0359 (10.1918)	2.6233 (11.5690)	3.2757 (12.9278)	3.9859 (14.2605)	4.7508 (15.5688)	5.5692 (16.8566)	6.4405 (18.1272)
$n=4$	2.8708 (12.1025)	3.5614 (13.4799)	4.3369 (14.8752)	5.1871 (16.2681)	6.1017 (17.6440)	7.0741 (18.9980)	8.1017 (20.3310)	9.1836 (21.6460)

Table 2.22 The q parameters and cut-on frequencies for aspect-ratio $D_2/D_1 = 0.97$ ($e = 0.2431$, $\xi_0 = 2.0923$)

Even Modes $\left.\dfrac{\mathrm{d}}{\mathrm{d}\xi}Ce_m\left(\xi,q_{m,n}\right)\right|_{\xi=\xi_0}=0$	$m=0$	$m=1$	$m=2$	$m=3$	$m=4$	$m=5$	$m=6$	$m=7$
$n=1$	0	0.0502	0.1418	0.2687	0.4305	0.6267	0.8567	1.1203
	(0)	(1.8434)	(3.0984)	(4.2644)	(5.3978)	(6.5126)	(7.6147)	(8.7076)
$n=2$	0.2240	0.4263	0.6818	0.9771	1.3112	1.6844	2.0961	2.5459
	(3.8938)	(5.3714)	(6.7933)	(8.1323)	(9.4204)	(10.6771)	(11.9109)	(13.1266)
$n=3$	0.7538	1.0952	1.5018	1.9548	2.4458	2.9764	3.5472	4.1579
	(7.1426)	(8.6097)	(10.0818)	(11.5025)	(12.8662)	(14.1933)	(15.4946)	(16.7755)
$n=4$	1.5922	2.0652	2.6149	3.2232	3.8717	4.5575	5.2835	6.0501
	(10.3809)	(11.8228)	(13.3034)	(14.7699)	(16.1877)	(17.5631)	(18.9102)	(20.2357)
Odd Modes $\left.\dfrac{\mathrm{d}}{\mathrm{d}\xi}Se_m\left(\xi,\bar{q}_{m,n}\right)\right|_{\xi=\xi_0}=0$	$m=1$	$m=2$	$m=3$	$m=4$	$m=5$	$m=6$	$m=7$	$m=8$
$n=1$	0.0531	0.1421	0.2687	0.4305	0.6267	0.8567	1.1203	1.4171
	(1.8957)	(3.1016)	(4.2646)	(5.3978)	(6.5126)	(7.6147)	(8.7076)	(9.7934)
$n=2$	0.4404	0.6856	0.9776	1.3112	1.6844	2.0961	2.5459	3.0331
	(5.4597)	(6.8118)	(8.1344)	(9.4206)	(10.6771)	(11.9109)	(13.1266)	(14.3278)
$n=3$	1.1295	1.5167	1.9583	2.4463	2.9765	3.5472	4.1579	4.8081
	(8.7433)	(10.1318)	(11.5127)	(12.8675)	(14.1934)	(15.4946)	(16.7755)	(18.0394)
$n=4$	2.1262	2.6506	3.2363	3.8745	4.5579	5.2835	6.0501	6.8573
	(11.9962)	(13.3939)	(14.8000)	(16.1936)	(17.5639)	(18.9103)	(20.2357)	(21.5433)

Table 2.23 The q parameters and cut-on frequencies for aspect-ratio $D_2/D_1 = 0.98$ ($e = 0.1990$, $\xi_0 = 2.2976$)

Even Modes $\left.\dfrac{\mathrm{d}}{\mathrm{d}\xi}Ce_m\left(\xi,q_{m,n}\right)\right|_{\xi=\xi_0}=0$	$m=0$	$m=1$	$m=2$	$m=3$	$m=4$	$m=5$	$m=6$	$m=7$
$n=1$	0	0.0336	0.0942	0.1783	0.2856	0.4157	0.5683	0.7432
	(0)	(1.8427)	(3.0841)	(4.2433)	(5.3710)	(6.4801)	(7.5767)	(8.6641)
$n=2$	0.1484	0.2841	0.4534	0.6486	0.8701	1.1176	1.3907	1.6891
	(3.8721)	(5.3574)	(6.7671)	(8.0942)	(9.3747)	(10.6248)	(11.8523)	(13.0618)
$n=3$	0.4984	0.7292	1.0001	1.2988	1.6236	1.9753	2.3539	2.7590
	(7.0955)	(8.5825)	(10.0507)	(11.4541)	(12.8063)	(14.1255)	(15.4198)	(16.6939)
$n=4$	1.0506	1.3729	1.7423	2.1445	2.5718	3.0256	3.5068	4.0152
	(10.3014)	(11.7762)	(13.2662)	(14.7179)	(16.1176)	(17.4820)	(18.8209)	(20.1390)
Odd Modes $\left.\dfrac{\mathrm{d}}{\mathrm{d}\xi}Se_m\left(\xi,\bar{q}_{m,n}\right)\right|_{\xi=\xi_0}=0$	$m=1$	$m=2$	$m=3$	$m=4$	$m=5$	$m=6$	$m=7$	$m=8$
$n=1$	0.0349	0.0942	0.1783	0.2856	0.4157	0.5683	0.7432	0.9401
	(1.8772)	(3.0855)	(4.2434)	(5.3710)	(6.4801)	(7.5767)	(8.6641)	(9.7445)
$n=2$	0.2904	0.4545	0.6487	0.8701	1.1176	1.3907	1.6891	2.0123
	(5.4157)	(6.7754)	(8.0948)	(9.3748)	(10.6248)	(11.8523)	(13.0618)	(14.2569)
$n=3$	0.7444	1.0049	1.2996	1.6237	1.9753	2.3539	2.7590	3.1902
	(8.6715)	(10.0748)	(11.4573)	(12.8066)	(14.1255)	(15.4198)	(16.6939)	(17.9512)
$n=4$	1.4006	1.7548	2.1474	2.5722	3.0257	3.5068	4.0152	4.5506
	(11.8943)	(13.3136)	(14.7279)	(16.1188)	(17.4821)	(18.8209)	(20.1390)	(21.4396)

Table 2.24 The q parameters and cut-on frequencies for aspect-ratio $D_2/D_1 = 0.99$ ($e = 0.1411$, $\xi_0 = 2.6467$)

Even Modes $\left.\dfrac{d}{d\xi}Ce_m\left(\xi,q_{m,n}\right)\right\|_{\xi=\xi_0}=0$	$m=0$	$m=1$	$m=2$	$m=3$	$m=4$	$m=5$	$m=6$	$m=7$
$n=1$	0 (0)	0.0169 (1.8419)	0.0469 (3.0693)	0.0887 (4.2222)	0.1421 (5.3442)	0.2068 (6.4478)	0.2828 (7.5389)	0.3697 (8.6209)
$n=2$	0.0738 (3.8514)	0.1421 (5.3441)	0.2259 (6.7381)	0.3228 (8.0551)	0.4329 (9.3287)	0.5561 (10.5724)	0.6920 (11.7936)	0.8404 (12.9971)
$n=3$	0.2475 (7.0531)	0.3644 (8.5582)	0.4989 (10.0143)	0.6467 (11.4016)	0.8081 (12.7448)	0.9830 (14.0567)	1.1713 (15.3443)	1.3729 (16.6118)
$n=4$	0.5208 (10.2312)	0.6854 (11.7377)	0.8702 (13.2253)	1.0686 (14.6560)	1.2804 (16.0425)	1.5060 (17.3985)	1.7453 (18.7299)	1.9982 (20.0410)
Odd Modes $\left.\dfrac{d}{d\xi}Se_m\left(\xi,\bar{q}_{m,n}\right)\right\|_{\xi=\xi_0}=0$	$m=1$	$m=2$	$m=3$	$m=4$	$m=5$	$m=6$	$m=7$	$m=8$
$n=1$	0.0172 (1.8590)	0.0469 (3.0697)	0.0887 (4.2222)	0.1421 (5.3442)	0.2068 (6.4478)	0.2828 (7.5389)	0.3697 (8.6209)	0.4677 (9.6958)
$n=2$	0.1436 (5.3730)	0.2260 (6.7402)	0.3228 (8.0551)	0.4329 (9.3287)	0.5561 (10.5724)	0.6920 (11.7936)	0.8404 (12.9971)	1.0012 (14.1862)
$n=3$	0.3682 (8.6024)	0.4996 (10.0207)	0.6468 (11.4020)	0.8081 (12.7448)	0.9830 (14.0567)	1.1713 (15.3443)	1.3729 (16.6118)	1.5874 (17.8628)
$n=4$	0.6924 (11.7973)	0.8720 (13.2391)	1.0688 (14.6573)	1.2804 (16.0426)	1.5060 (17.3985)	1.7453 (18.7299)	1.9982 (20.0410)	2.2645 (21.3348)

Table 2.25 The q parameters and cut-on frequencies for aspect-ratio $D_2/D_1 = 0.999$ ($e = 0.0447$, $\xi_0 = 3.8002$)

Even Modes $\left.\dfrac{d}{d\xi}Ce_m\left(\xi,q_{m,n}\right)\right\|_{\xi=\xi_0}=0$	$m=0$	$m=1$	$m=2$	$m=3$	$m=4$	$m=5$	$m=6$	$m=7$
$n=1$	0 (0)	0.0017 (1.8413)	0.0047 (3.0558)	0.0088 (4.2033)	0.0141 (5.3202)	0.0206 (6.4188)	0.0281 (7.5050)	0.0368 (8.5821)
$n=2$	0.0073 (3.8336)	0.0142 (5.3327)	0.0225 (6.7095)	0.0321 (8.0192)	0.0431 (9.2870)	0.0554 (10.5251)	0.0689 (11.7408)	0.0837 (12.9389)
$n=3$	0.0246 (7.0191)	0.0364 (8.5384)	0.0497 (9.9744)	0.0644 (11.3516)	0.0805 (12.6882)	0.0979 (13.9942)	0.1166 (15.2758)	0.1367 (16.5376)
$n=4$	0.0518 (10.1786)	0.0685 (11.7089)	0.0868 (13.1768)	0.1064 (14.5931)	0.1275 (15.9720)	0.1499 (17.3215)	0.1738 (18.6468)	0.1989 (19.9518)
Odd Modes $\left.\dfrac{d}{d\xi}Se_m\left(\xi,\bar{q}_{m,n}\right)\right\|_{\xi=\xi_0}=0$	$m=1$	$m=2$	$m=3$	$m=4$	$m=5$	$m=6$	$m=7$	$m=8$
$n=1$	0.0017 (1.8430)	0.0047 (3.0558)	0.0088 (4.2033)	0.0141 (5.3202)	0.0206 (6.4188)	0.0281 (7.5050)	0.0368 (8.5821)	0.0466 (9.6522)
$n=2$	0.0142 (5.3355)	0.0225 (6.7095)	0.0321 (8.0192)	0.0431 (9.2870)	0.0554 (10.5251)	0.0689 (11.7408)	0.0837 (12.9389)	0.0997 (14.1226)
$n=3$	0.0365 (8.5428)	0.0497 (9.9745)	0.0644 (11.3516)	0.0805 (12.6882)	0.0979 (13.9942)	0.1166 (15.2758)	0.1367 (16.5376)	0.1580 (17.7829)
$n=4$	0.0686 (11.7149)	0.0868 (13.1770)	0.1064 (14.5931)	0.1275 (15.9721)	0.1499 (17.3215)	0.1738 (18.6468)	0.1989 (19.9518)	0.2254 (21.2397)

Table 2.26 The Roots of the derivative of Bessel function of the first kind: non-dimensional cut-on frequencies in a circular cylindrical waveguide $D_2 = D_1 = D_0$ $(e = 0, \quad \xi_0 \rightarrow \infty)$

$\alpha_{m,n}$	$m = 0$	$m = 1$	$m = 2$	$m = 3$	$m = 4$	$m = 5$	$m = 6$	$m = 7$	$m = 8$
$n = 0$	0	1.8412	3.0542	4.2012	5.3176	6.4156	7.5013	8.5778	9.6474
$n = 1$	3.8317	5.3314	6.7061	8.0152	9.2824	10.5199	11.7349	12.9324	14.1155
$n = 2$	7.0156	8.5363	9.9695	11.3459	12.6819	13.9872	15.2682	16.5294	17.7740
$n = 3$	10.1735	11.7060	13.1704	14.5858	15.9641	17.3128	18.6374	19.9419	21.2291
$n = 4$	13.3237	14.8636	16.3475	17.7887	19.1960	20.5755	21.9317	23.2681	24.5872
$n = 5$	16.4706	18.0155	19.5129	20.9725	22.4010	23.8036	25.1839	26.5450	27.8893
$n = 6$	19.6159	21.1644	22.6716	24.1449	25.5898	27.0103	28.4098	29.7907	31.1553
$n = 7$	22.7601	24.3113	25.8260	27.3101	28.7678	30.2028	31.6179	33.0152	34.3966
$n = 8$	25.9037	27.4571	28.9777	30.4703	31.9385	33.3854	34.8134	36.2244	37.6201

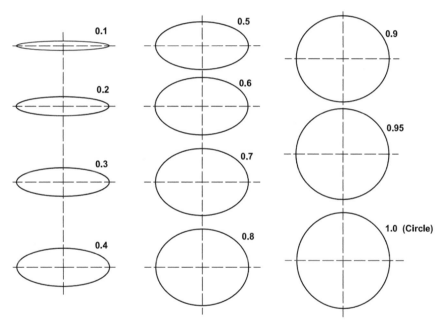

Fig. 2.4 Family of ellipses of constant major-axis showing a gradual transition from a highly eccentric to zero eccentric (circle) section. The aspect-ratio value is shown adjacent to each ellipse. The diameter of the circle is the same as the major-axis

The resonance frequency of the first few even and odd transverse modes of a rigid-wall elliptical duct/chamber for a given eccentricity (aspect-ratio) reported in the previous papers [31, 36, 40] was found to be the same as those computed using the numerical root-finding algorithm described in Sect. 2.5. Additionally, the accuracy of the tabulated values and the corresponding mode shapes presented in an ensuing section were verified by comparing them with counterpart values computed numerically using a set of FE codes through the solution of an algebraic eigenvalue problem.

2.7 Resonance Frequency Variation with Aspect-Ratio, Interpolating Polynomials and Mode Shapes

2.7.1 Circumferential Modes of Even and Odd Type

Figure 2.5 shows the variation of resonance frequency of the first few circumferential modes with aspect-ratio—dots ● and cross × indicate exact tabulated values of even and odd modes, respectively, while the broken and solid lines denote a highly accurate least-squares interpolating polynomial of 10th degree, respectively, given by Eq. (2.68).

$$\left(k_0 \frac{D_1}{2}\right)\Bigg|_{m,n} = a_0 + a_1\beta + a_2\beta^2 + a_3\beta^3 + a_4\beta^4 + a_5\beta^5$$
$$+ a_6\beta^6 + a_7\beta^7 + a_8\beta^8 + a_9\beta^9 + a_{10}\beta^{10}, \qquad (2.68)$$

where $\beta = \left(\frac{D_2}{D_1} - 0.62391\right)/0.30319$ is the centered and scaled aspect-ratio variable, and a_j's denote optimized polynomial coefficients shown up to five significant places of decimal in Table 2.27 for the first four even and odd circumferential modes.

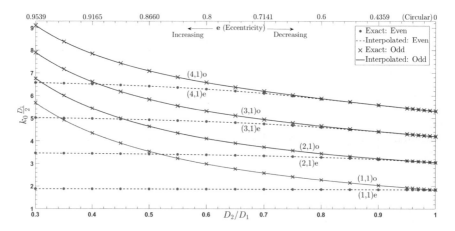

Fig. 2.5 Variation of non-dimensional resonance frequency of the first few circumferential modes of the even and odd types, with aspect-ratio D_2/D_1. A high-order least-squares interpolating polynomial is fitted to the exact values of the even and odd modes which are shown by ● and ×, respectively

Table 2.27 Coefficients of a 10th degree least-squares fit polynomial for interpolating the non-dimensional resonance frequency of the first few even and odd circumferential modes

Coefficients	$(1, 1)e$	$(1, 1)o$	$(2, 1)e$	$(2, 1)o$	$(3, 1)e$	$(3, 1)o$	$(4, 1)e$	$(4, 1)o$
a_0	1.8669	2.8744	3.3768	4.0014	4.8333	5.2150	6.2496	6.4746
a_1	-0.017773	-1.2951	-0.12289	-1.2369	-0.28853	-1.2036	-0.51505	-1.2213
a_2	-3.1492×10^{-3}	0.6792	-0.056177	0.69431	-0.15815	0.65005	-0.28066	0.56535
a_3	6.3732×10^{-4}	-0.39456	-0.025582	-0.39773	-0.067349	-0.41781	-0.045746	-0.42923
a_4	3.3307×10^{-4}	-0.22331	-0.014621	-0.22703	-2.1843×10^{-3}	-0.22618	0.088644	-0.21618
a_5	-2.4822×10^{-4}	0.2084	-6.1031×10^{-3}	0.21158	0.03024	0.21228	0.063392	0.21209
a_6	-3.4240×10^{-4}	0.6157	3.1287×10^{-4}	0.6143	0.01613	0.60934	-0.020027	0.6072
a_7	1.8513×10^{-4}	-0.25795	2.5050×10^{-3}	-0.25845	-3.8624×10^{-3}	-0.25671	-0.031391	-0.25223
a_8	1.7423×10^{-4}	-0.3832	1.0787×10^{-3}	-0.38236	-4.7138×10^{-3}	-0.38044	-2.2914×10^{-3}	-0.37943
a_9	-4.6580×10^{-5}	0.062519	-3.2266×10^{-5}	0.062488	-4.0512×10^{-4}	0.06199	6.1710×10^{-3}	0.060878
a_{10}	-3.4661×10^{-5}	0.096933	-6.9002×10^{-5}	0.096798	2.3061×10^{-4}	0.096486	1.7144×10^{-3}	0.096187

Equation (2.68) in conjunction with Table 2.27 may be used to accurately compute the resonance frequencies of the first four even and odd circumferential modes up to four decimal places for aspect-ratio between $D_2/D_1 = [0.3, 1]$ sufficient to account for elliptical silencers used in most engineering applications. A least-squares interpolating polynomial may likewise be also developed for still higher-order even and odd circumferential modes.

Figure 2.5 shows that the resonance frequency graphs of even and odd circumferential modes of a given order m coalesce in the limit $D_2/D_1 \rightarrow 1$ (circular) as anticipated. In this case, the angular and radial Mathieu function behavior tends to that of trigonometric and Bessel functions of the first kind, respectively, wherein the even-even and odd-even modes coalesce, and the even–odd and odd-odd modes coalesce to the circumferential modes of the circular chamber. For example, the $(1, 1)e$ and $(1, 1)o$ modes of the elliptical duct approach the $(1, 0)$ mode of the circular duct, the $(2, 1)e$ and $(2, 1)o$ modes approach the $(2, 0)$ mode, and so on. Therefore, for elliptical sections having very small eccentricity, it is equally easy to excite the even and odd modes of a given order. The frequency graphs, however, bifurcate with decreasing aspect-ratio, i.e., with increasing eccentricity. It is observed that resonance frequency of odd circumferential modes increases significantly with eccentricity while comparatively those of even modes do not increase as much—note the mildly convex nature and concave nature of the even and odd mode frequency graphs, respectively. In particular, the cut-on frequency graph of the first transverse mode, i.e., the $(1, 1)e$ mode graph, is nearly a horizontal line—for $D_2/D_1 = 1$, $(0.5k_0 D_1)|_{(1,1)e} = 1.8412$—whereas for $D_2/D_1 = 0.3$, $(0.5k_0 D_1)|_{(1,1)e} = 1.8818$ signifying only about 2% increase across the aspect-ratio range. Since the resonance frequency of $(1, 1)e$ mode does not change significantly with aspect-ratio for a family of ellipses with a constant major-axis D_1, an average value can be computed and is given by $0.5k_0 D_1 \approx 1.86$. In other words, the average cut-off frequency is given by $f_0 = 0.5933 \frac{c_0}{D_1}$—incidentally, this compares well with that of a circular chamber for which $f_0 = 0.5861 \frac{c_0}{D_0}$ or $0.5k_0 D_0 = 1.84$ where $D_0 = D_1$. Note that the $(1, 1)e$ mode is the first higher-order transverse mode, therefore, below the cut-on frequency of this mode, the wave propagation is purely planar. On the other hand, the resonance frequency of $(1, 1)o$ mode equals 5.6809 when $D_2/D_1 = 0.3$, thereby indicating about 209% increase! Additionally, it is noted that the frequency graph of an odd mode intersects those of higher-order even modes in the small aspect-ratio range signifying that for certain elliptical cross-sections, it is equally to excite a lower-order odd mode and a higher-order even mode—this property is useful in analysis and design of short end-chamber elliptical mufflers (Chap. 4). Another noteworthy observation is that the even and odd mode graphs of higher-order circumferential modes tend to coalesce at a progressively smaller aspect-ratio. Nevertheless, even modes, in general, are easier to be excited, or in other words, they propagate at a lower frequency in comparison with their odd counterparts.

2.7.1.1 Mode Shapes and Pressure Nodal Hyperbolas

Figure 2.6a–c shows the first three circumferential modes of the even type, i.e., $Ce_m(\xi, q_{m,1})ce_m(\eta, q_{m,1})$ where $m = 1, 2, 3 \ldots$ respectively, while Figs. 2.6d–f shows the first three circumferential modes of the odd type, i.e., $Se_m(\xi, \overline{q}_{m,1})se_m(\eta, \overline{q}_{m,1})$ where $m = 1, 2, 3 \ldots$ respectively, of a highly eccentric elliptical section with $D_2/D_1 = 0.5$, i.e., $e = 0.8660$. (The (0, 1) mode of an ellipse is the plane wave mode with uniform acoustic pressure distribution across the waveguide cross-section.) The even circumferential mode shapes presented in Figs. 2.6a–c demonstrate a standing-wave pattern predominantly along the major-axis with regions of alternate phase which are demarcated by pressure nodal hyperbolas. For example, the $(1, 1)e$ mode shape shows that the acoustic pressure field over one-half of the ellipse is in opposite phase with respect to the other half, and the two regions are separated by pressure nodal hyperbolas co-incident with the minor-axis, i.e., a hyperbola with $\eta_{\text{nodal}} = \pi/2$ in $0 \leq \eta \leq \pi$ interval and $\eta_{\text{nodal}} = 3\pi/2$ in $\pi \leq \eta \leq 2\pi$ interval. Therefore, by locating one of the ports along the minor-axis or at the ellipse center, the $(1, 1)e$ mode can be suppressed. Similarly, the $(2, 1)e$ mode exhibits three alternate phase regions where nodal hyperbolas are given by $\eta_{\text{nodal}} = \{1.0594, 2.0822\}$ in $0 \leq \eta \leq \pi$ interval, while the $(3, 1)e$ mode exhibits four alternate phase regions with $\eta_{\text{nodal}} = \{0.8227, \pi/2, 2.3189\}$ in $0 \leq \eta \leq \pi$ interval, and so on for higher-order even circumferential modes. In general, an even circumferential mode of order m, i.e., $Ce_m(\xi, q_{m,1})ce_m(\eta, q_{m,1})$, has m confocal nodal hyperbolas because the function angular Mathieu function $ce_m(\eta, q_{m,1})$ has m simple ordered zeros in $0 \leq \eta < \pi$ interval denoted by $\eta_{\text{noda}} = (\eta_1 < \eta_2 < \eta_3 < \cdots < \eta_m)$ which, in general, can be computed by numerically solving $ce_m(\eta, q_{m,1}) = 0$. Note that the pressure nodal hyperbolas always intersect the major-axis at a point between the ellipse center $\mathbf{O}$ and a focus $\mathbf{F}$ (refer to Fig. 2.1), i.e., on the semi-interfocal distance. Therefore, the point of intersection of the major-axis and $\eta_{(2,1)e}$ nodal hyperbola yields the required point where one of the ports can be centered to suppress the $(2, 1)e$ mode [36]. (Table 4.1 presents numerically computed pressure nodal hyperbolas $\eta_{(2,1)e}$ for a range of aspect-ratio—this data is used to design short elliptical camber mufflers to achieve a broadband attenuation performance.) Furthermore, the even-odd circumferential modes, i.e., $Ce_1(\xi, q_{1,1})ce_1(\eta, q_{1,1})$, $Ce_3(\xi, q_{3,1})ce_3(\eta, q_{3,1})$, $Ce_5(\xi, q_{5,1})ce_5(\eta, q_{5,1}), \ldots$ modes, always have a nodal hyperbola co-incident along the minor-axis, i.e., $\eta_{\text{nodal}} = \pi/2$ and $\eta_{\text{nodal}} = 3\pi/2$. In other words, the acoustic pressure field for such modes is anti-symmetric about the minor-axis. On the other hand, the even-even circumferential modes, i.e., $Ce_2(\xi, q_{2,1})ce_2(\eta, q_{2,1})$, $Ce_4(\xi, q_{4,1})ce_4(\eta, q_{4,1})$, $Ce_0(\xi, q_{0,1})ce_0(\eta, q_{0,1})$ modes, do not have such a nodal hyperbola; rather, they are symmetric about the minor-axis. Both even-even and even–odd circumferential modes are symmetric about the major-axis.

The odd circumferential modes always exhibit a pressure nodal hyperbolas co-incident with the major-axis, i.e., a hyperbola with $\eta_{\text{nodal}} = 0$ in $0 \leq \eta < \pi$ and $\eta_{\text{nodal}} = \pi$ in $\pi \leq \eta < 2\pi$ interval which may be verified by substituting $\eta = 0$ or $\eta = \pi$ in Eqs. (2.20) and (2.21). A nodal line along the major-axis in Figs. 2.6d–f shows the nodal hyperbolas signifying that the odd circumferential modes are

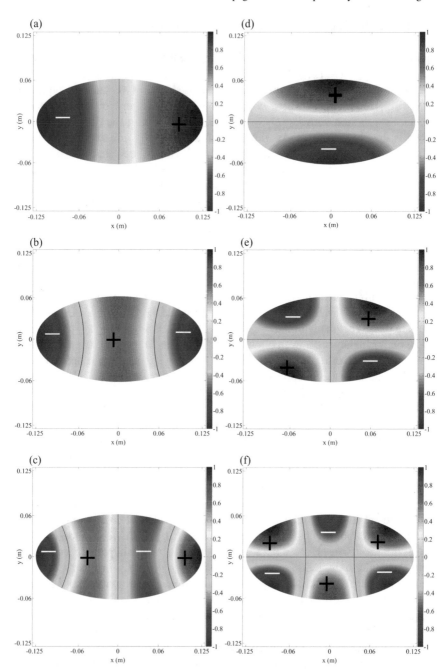

Fig. 2.6 Mode shapes of the first three even and odd circumferential modes of the ellipse having aspect-ratio $D_2/D_1 = 0.5$. Parts **a–c** show the $(1, 1)e$, $(2, 1)e$ and $(3, 1)e$ mode shapes, respectively, while parts **d–f** show the $(1, 1)o$, $(2, 1)o$ and $(3, 1)o$ mode shapes, respectively. These mode shapes are also schematically presented in Denia et al. [36] and Hong and Kim [39]

always anti-symmetric *w.r.t.* the major-axis. Specifically, for the (1, 1)*o* mode, it signifies that the acoustic pressure fluctuations on either sides of the major-axis are in opposite phase. The (1, 1)*o* mode and the higher odd–odd circumferential modes, however, are symmetric about the minor-axis. The odd–even circumferential modes, on other hand, always have $\eta_{\text{nodal}} = \pi/2$ and $\eta_{\text{nodal}} = 3\pi/2$ signifying that these modes are also anti-symmetric about the minor-axis. As a consequence, the acoustic pressure field of the (2, 1)*o* mode exhibits quadrants with alternate phase. Similar to the even circumferential modes, the odd circumferential modes of order m, i.e., $Se_m(\xi, \overline{q}_{m,1})se_m(\eta, \overline{q}_{m,1})$, also have m confocal nodal hyperbolas because $se_m(\eta, \overline{q}_{m,1})$ has m simple ordered zeros in $0 \leq \eta < \pi$ interval (computed by solving $se_m(\eta, \overline{q}_{m,1}) = 0$) which, in general, divides the ellipse in sector-like regions having alternate phase, see, for example, the (3, 1)*o* mode.

The system of confocal nodal hyperbolas arises in both even and odd circumferential modes because the angular Mathieu function Eq. (2.13) belongs to the class of Sturm–Liouville (S-L) problem as they can be put in the following form [45, 46]

$$\frac{\mathrm{d}}{\mathrm{d}\eta}\left[\frac{\mathrm{d}p_\eta}{\mathrm{d}\eta}\right] - (2q\cos 2\eta)p_\eta = -a_m p_\eta, \tag{2.69}$$

where $a_m^{'s}$ denote a sequence of real eigenvalues that were found by solving the eigenvalue problem in Eqs. (2.26–2.29). Here, for an eigenvalue a_m, the eigenfunctions $ce_m(\eta, q_{m1})$ or $se_m(\eta, \overline{q}_{m,1})$ have exactly $m - 1$ zeros in $0 \leq \eta < \pi$ interval [5].

For the circular chamber, the circumferential mode shapes given by $J_m\left(\alpha_{m,0}\frac{r}{R_0}\right)\cos m\theta$ and $J_m\left(\alpha_{m,0}\frac{r}{R_0}\right)\sin m\theta$ are essentially the same except $\pi/2$ phase difference. The former is a counterpart of the even circumferential elliptical mode, while the latter is the counterpart of the odd circumferential elliptical mode— as noted earlier, the even and odd modes coalesce as the ellipse approaches the degenerate case of a circle. The first three circumferential modes $J_m\left(\alpha_{m,0}\frac{r}{R_0}\right)\cos m\theta$ shown in Fig. 2.7a–c demonstrate that the confocal nodal hyperbolas collapse to nodal diameters for the circular case which can be found by setting $\cos m\theta = 0$. In $0 \leq \theta < \pi$ interval, the nodal diameters (or radii) are given by $\theta_{\text{nodal}} = (2n+1)\frac{\pi}{2m}$, where $n = 0, 1, \ldots, m - 1$ and they divide the circle in regions having alternate phase.

2.7.2 Radial Modes and Cross-Modes of Even and Odd Type

Figure 2.8 shows the variation of tabulated resonance frequency values of the first radial and cross-modes of even type ●and cross-modes of odd type ×, while the broken and solid lines, respectively, denote the corresponding 10th degree least-squares interpolating polynomial. Equation (2.68) along with Table 2.28 may also be used to accurately compute up to four decimal places, the resonance frequencies

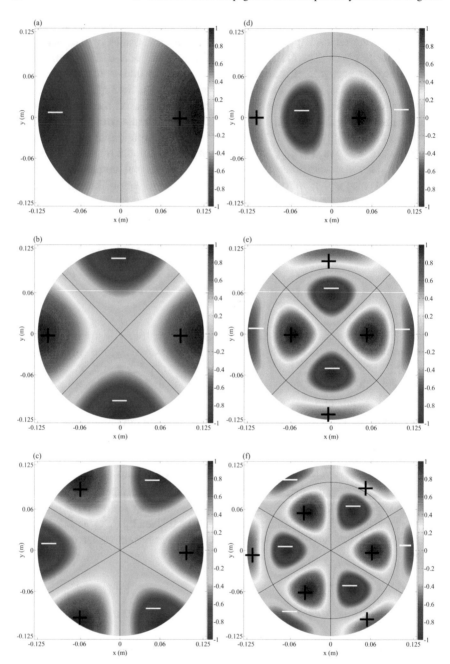

Fig. 2.7 **a–c** Mode shapes of the first three circumferential modes of the circle: **a** $(1, 0)$ mode, **b** $(2, 0)$ mode and **c** $(3, 0)$ mode. **d–f** Mode shapes of the cross-modes: **d** $(1, 1)$ mode, **e** $(2, 1)$ mode and **f** $(3, 1)$ mode of the circle

Table 2.28 Coefficients of a 10th degree least-squares fit polynomial for interpolating the non-dimensional resonance frequency of the first radial mode and first few cross-modes

Coefficients	$(0, 2)e$	$(1, 2)e$	$(1, 2)o$	$(2, 2)e$	$(2, 2)o$	$(3, 2)e$	$(3, 2)o$
a_0	5.5135	6.6804	8.0481	7.9320	9.1848	9.2540	10.375
a_1	−2.3924	−2.2203	−3.6046	−1.9965	−3.4789	−1.7544	−3.3366
a_2	1.3964	1.5043	2.0495	1.6040	2.1006	1.6153	2.1361
a_3	−0.7698	−0.72206	−1.1880	−0.75705	−1.1733	−0.91068	−1.1807
a_4	−0.42696	−0.44107	−0.65829	−0.5492	−0.65991	−0.59604	−0.68258
a_5	0.42212	0.38482	0.62502	0.35106	0.62029	0.45254	0.60738
a_6	1.2304	1.2146	1.8456	1.2619	1.8431	1.2393	1.8437
a_7	−0.51777	−0.51113	−0.77285	−0.48270	−0.77314	−0.58589	−0.77075
a_8	−0.76759	−0.76167	−1.1496	−0.77400	−1.1493	−0.78043	−1.1504
a_9	0.12497	0.12529	0.18734	0.11189	0.18753	0.14918	0.18603
a_{10}	0.19396	0.19362	0.29079	0.19076	0.29082	0.20251	0.29046

of the $(0, 2)e$ radial mode and $(1, 2)e$, $(1, 2)o$, $(2, 2)e$, $(2, 2)o$, $(3, 2)e$ and $(3, 2)o$ cross-modes of the even and odd types for $D_2/D_1 = [0.3, 1]$. To accurately compute the resonance frequency of still higher-order radial and cross-modes for any aspect-ratio, appropriate values in Tables presented in Sect. 2.5 may likewise be interpolated using a least-squares polynomial fit.

While the even and odd cross-mode graphs coalesce in the limit $D_2/D_1 \rightarrow 1$, the resonance frequencies increase with decreasing aspect-ratio as may also be noted from the concave nature of the graphs shown in Fig. 2.8. (Similar comments also hold good for frequency graphs of higher-order radial modes $(0, n)$ and cross-modes

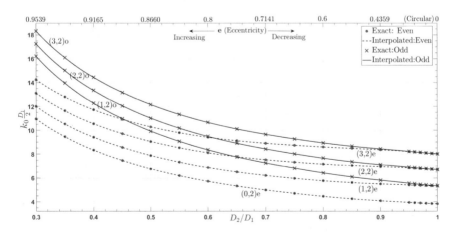

Fig. 2.8 Variation of non-dimensional resonance frequency of the first radial mode and first few cross-modes of the even and odd type, with aspect-ratio D_2/D_1. Similar to Fig. 2.5, a least-squares interpolating polynomial is fitted to the exact values

(m, n) where $n = 3, 4, 5, \ldots$) In fact, at lower aspect-ratios, the resonance frequency of cross-modes is significantly larger than that of circumferential modes of the same order shown in Fig. 2.5 signifying that for a highly eccentric section, it is relatively much more difficult to excite the cross as well as radial modes in which case the ellipse is long and narrow resembling a ribbon whose width is $2\,h$, i.e., the interfocal distance [5].

From the point-of-view of design and analysis of elliptical mufflers, it is important to mention that the cut-on frequency of the first two even circumferential modes, i.e., the $(1, 1)e$ and $(2, 1)e$ modes, is lower than that of the first radial mode $(0, 2)e$ for all aspect-ratios. Indeed, for highly eccentric elliptical sections, the cut-on frequencies of still higher-order even circumferential modes and first few odd circumferential modes are also lower than that of the $(0, 2)e$ mode, compare the frequency graphs in Figs. 2.5 and 2.8.

2.7.2.1 Mode Shapes and Pressure Nodal Ellipses and Hyperbolas

Figure 2.9a, c, e shows the first three radial modes $Ce_0(\xi, q_{0,2})ce_0(\eta, q_{0,2})$, $Ce_0(\xi, q_{0,3})ce_0(\eta, q_{0,3})$, and $Ce_0(\xi, q_{0,4})ce_0(\eta, q_{0,4})$, respectively, of an elliptical chamber with $D_2/D_1 = 0.5$, while Fig. 2.9b, d, f shows the first three radial modes of the circular chamber given by $J_0\left(\alpha_{0,n}\frac{r}{R_0}\right)$ where $n = 1, 2$ and 3, respectively. The radial mode shapes presented in Fig. 2.9a–c exhibit confocal annular regions of alternating phase demarcated by pressure nodal ellipses. For example, the $(0, 2)e$ or first radial mode shape shows that the acoustic pressure field over the inner elliptical region is in opposite phase with respect to that over the annular elliptical region, and the two regions are separated by a pressure nodal ellipse ξ_{nodal} given by $\xi_{(0,2)e} = \xi_1 = 0.2913$. Therefore, by centering one of the ports at the point of intersection of the minor-axis and the pressure nodal ellipse of the $(0, 2)e$ mode, this mode can be suppressed in addition to the suppression of even–odd and odd–even modes which effectively increases the broadband attenuation range. This will be explained in a greater detail and illustrated for a short muffler configuration in the ensuing chapters. The $(0, 3)e$ or second radial mode exhibits three alternate phase regions separated by two nodal ellipses at $\xi_{(0,3)e} = \{\xi_1, \xi_2\} \approx \{0.1401, 0.4201\}$, and so on. In general, an elliptical radial mode $Ce_0(\xi, q_{0,n})ce_0(\eta, q_{0,n})$ has $n - 1$ confocal nodal ellipse(s) which separate the radial mode in n confocal regions of alternating phase. This is because the modified Mathieu function $Ce_0(\xi, q_{0,n})$ has exactly $n - 1$ simple ordered zeros in the interval $\xi = \left[0, \xi_0 = \cosh^{-1}(1/e)\right]$ denoted by $\xi_{\text{nodal}} = (\xi_1 < \xi_2 < \xi_3 < \cdots < \xi_{n-1})$ which are computed by numerically solving $Ce_0(\xi, q_{0,n}) = 0$. (Table 4.1 also presents numerically computed pressure nodal ellipses $\xi_{(0,2)e}$ corresponding to the first radial mode for a range of aspect-ratio—this data is used for double-tuning the acoustic performance of short chambers.) Note that the system of confocal nodal ellipses arises for both radial and cross-modes because the modified Mathieu functions also belong to the class of S-L problem given by

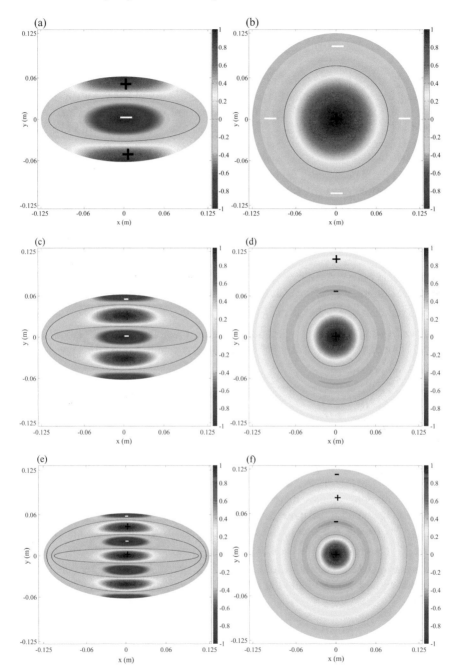

Fig. 2.9 Parts **a**, **c**, **e**: radial modes of the elliptical chamber with $D_2/D_1 = 0.5$: **a** $(0, 2)e$, **c** $(0, 3)e$ and **e** $(0, 4)e$ modes [36, 40]. Parts **b**, **d**, **f**: Radial modes of the circular chamber: **b** $(0, 1)$, **d** $(0, 2)$ and **f** $(0, 3)$ modes

$$\frac{d}{d\xi}\left[\frac{dp_\xi}{d\xi}\right] + (2q\cosh 2\xi)\,p_\xi = a_m\,p_\xi, \qquad (2.70)$$

where $a_m'^s$ denote a sequence of real eigenvalues as noted earlier. Therefore, corresponding to each a_m, the unique eigenfunction $Ce_m(\xi, q_{m,n})$ has exactly $n-1$ simple zeros in the interval $\xi = [0, \xi_0]$ which is observed from the radial mode shapes shown in Fig. 2.9a, c, e as well as from the cross-mode shapes shown in Fig. 2.10, see McLachlan [5].

When the elliptical cross-section tends to a circular cross-section, the nodal ellipses in parts (a), (c) and (e) of Fig. 2.9 degenerate to concentric nodal circle(s) as observed from parts (b), (d) and (f) which demonstrate the axisymmetric nature of the circular radial modes unlike its elliptical counterparts. The first circular radial mode $J_0\left(\alpha_{0,1}\frac{r}{R_0}\right)$ has only one nodal circle located at a radius $r = 0.6276R_0$ which implies that the inner circular region is vibrating with an opposite phase as compared to the concentric annular region. Similarly, the second and third circular radial modes have two and three nodal circles, respectively, thereby indicating concentric annular regions of alternating phase. Given that Bessel function belongs to the category of S-L problem, the solution of $J_0\left(\alpha_{0,n}\frac{r}{R_0}\right) = 0$ yields n simple roots given by $r_{\text{nodal}} = (r_1 < r_2 < r_3 < \cdots < r_n)$ which denote the nodal circles for the nth radial mode. Note that the same is true for computing the n nodal circles for a cross-mode $J_m\left(\alpha_{m,n}\frac{r}{R_0}\right)\cos m\theta$.

Figure 2.10a, c, e shows the first three cross-modes of the even type given by $Ce_1(\xi, q_{1,2})ce_1(\eta, q_{1,2})$, $Ce_2(\xi, q_{2,2})ce_2(\eta, q_{2,2})$, $Ce_3(\xi, q_{3,2})ce_3(\eta, q_{3,2})$, respectively, while Figure 2.10b, d, f shows the first three cross-modes of the odd type given by $Se_1(\xi, \bar{q}_{1,2})se_1(\eta, \bar{q}_{1,2})$, $Se_2(\xi, \bar{q}_{2,2})se_2(\eta, \bar{q}_{2,2})$, $Se_3(\xi, \bar{q}_{3,2})se_3(\eta, \bar{q}_{3,2})$, respectively, for the elliptical chamber with aspect-ratio $D_2/D_1 = 0.5$ The cross-mode shapes shown in Fig. 2.10 reveal the existence of both pressure nodal hyperbolas and a nodal ellipse which further divide the elliptical cross-section into smaller regions of alternating phase. In fact, parts (a–f) of Fig. 2.10 are similar to their counterpart circumferential mode shapes shown in Fig. 2.6 except the presence of a nodal ellipse. Similarly, the cross-modes $J_1\left(\alpha_{1,1}\frac{r}{R_0}\right)\cos\theta$, $J_2\left(\alpha_{2,1}\frac{r}{R_0}\right)\cos 2\theta$, $J_3\left(\alpha_{3,1}\frac{r}{R_0}\right)\cos 3\theta$ of the circular chamber shown in Fig. 2.7d–f, respectively, indicate the simultaneous presence of nodal diameters and circles which divide the circular cross-section in regions of alternating phase.

2.8 Family of Ellipses Having Equal Cross-Sectional Area

The discussion, so far, has focused on a family of ellipses which have a constant major-axis regardless of eccentricity, i.e., the diameter of a circular section is the

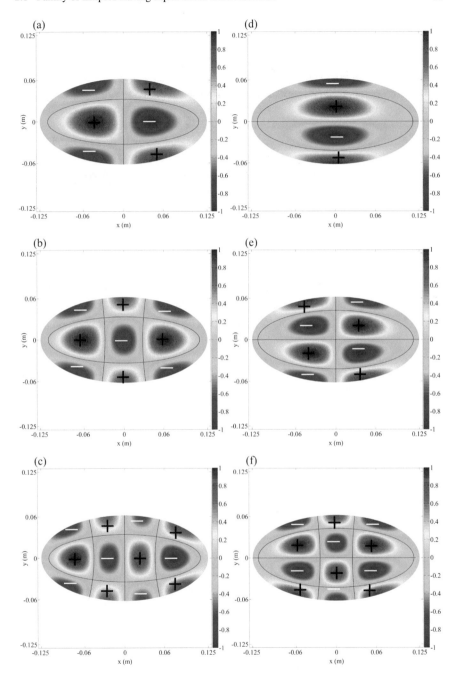

Fig. 2.10 Mode shapes of the first three even and odd cross-modes of the ellipse having $D_2/D_1 = 0.5$ parts **a–c** show the $(1, 2)e$, $(2, 2)e$ and $(3, 2)e$ mode shapes, respectively, while parts **d–f** show the $(1, 2)o$, $(2, 2)o$ and $(3, 2)o$ mode shapes, respectively [39]

same as that of a highly eccentric ellipse. However, in several applications where one must investigate the effect of departure from axisymmetry, the propagation conditions of higher-order modes are analyzed within a family of elliptical ducts having the same cross-sectional area [35]. This is because more meaningful conclusions can be drawn by comparing cylindrical acoustic cavities of the same volume because they will occupy approximately the same space in actual machinery like exhaust systems [39]. To this end, a circular duct of diameter D_{eq} is considered which is deformed into an elliptical duct in such a manner that it is elongated along a diameter and simultaneously compressed along the perpendicular direction. In order for the resulting elliptical duct to have the same cross-sectional area as its circular counterpart, the major-axis D_1 and minor-axis D_2 are related to D_{eq} by the following expression.

$$D_{eq} = \sqrt{D_1 D_2} = D_1 \sqrt{\frac{D_2}{D_1}} = D_1 \sqrt[4]{1 - e^2}, \qquad (2.70a\text{--}c)$$

where $D_2 = D_1\sqrt{1 - e^2}$ and as usual, e denotes the eccentricity. Now, for a circular duct, the non-dimensional resonance frequency of the (m, n) mode denoted by $\left(k_0 \frac{D_{eq}}{2}\right)_{m,n}$ is related to its elliptical counterpart by the following expression.

$$\left(k_0 \frac{D_{eq}}{2}\right)_{m,n} = \sqrt[4]{1 - e^2} \times \alpha_{m,n}^{\text{elliptical}} = \sqrt{\frac{D_2}{D_1}} \times \alpha_{m,n}^{\text{elliptical}}, \qquad (2.71)$$

where

$$\alpha_{m,n}^{\text{elliptical}} = \left(k_0 \frac{D_1}{2}\right)\bigg|_{(m,n)\text{ Even}} \qquad (2.72a)$$

or

$$\alpha_{m,n}^{\text{elliptical}} = \left(k_0 \frac{D_1}{2}\right)\bigg|_{(m,n)\text{Odd}}, \qquad (2.72b)$$

denotes the non-dimensional resonance frequency of an elliptical duct that may be obtained from Tables 2.1, 2.2, 2.3, 2.4, 2.5, 2.6, 2.7, 2.8, 2.9, 2.10, 2.11, 2.12, 2.13, 2.14, 2.15, 2.16, 2.17, 2.18, 2.19, 2.20, 2.21, 2.22, 2.23, 2.24, 2.25 for a given aspect-ratio. Equation (2.71) suggests that the resonance frequencies of a family of ellipses having the same cross-sectional area may be obtained by multiplying or scaling the non-dimensional values in Tables 2.1, 2.2, 2.3, 2.4, 2.5, 2.6, 2.7, 2.8, 2.9, 2.10, 2.11, 2.12, 2.13, 2.14, 2.15, 2.16, 2.17, 2.18, 2.19, 2.20, 2.21, 2.22, 2.23, 2.24, 2.25 with the factor $\sqrt{D_2/D_1}$, i.e., with the square-root of the aspect-ratio. In Lowson and Baskaran [35], a family of ellipses with the same cross-sectional area π was considered which yielded the following relations

$$h^2 = e^2 \left(1 - e^2\right)^{-1/2},$$
$$w_0^2 = s/e^2 = k^2 \left(1 - e^2\right)^{-1/2}, \text{ and}$$
$$s \equiv 4q,$$

(2.73–2.75)

where s and w_0 were new variables that were introduced, and w_0 was the same as the non-dimensional frequency. The root-bracketing and bisection methods were implemented to numerically determine the zeros S corresponding to the Neumann or rigid-wall condition at the elliptical boundary following which the transverse cut-on wavenumber k was determined for a given e. As mentioned previously, they documented the non-dimensional resonance frequencies of the first 15 orders of even and odd circumferential modes for a range of eccentricity which were found to be identical to those computed here.

Figure 2.11 shows the graphical variation w.r.t. the aspect-ratio of the non-dimensional resonance frequency of the first four even and odd circumferential modes and the first radial mode of a family of ellipses having equal cross-sectional area. The graphs for even circumferential modes in Fig. 2.11 exhibit a convex variation, but unlike Fig. 2.5, their slopes increase w.r.t. the aspect-ratio, i.e., the cut-on frequency of the higher-order even circumferential modes decreases for smaller aspect-ratio or more eccentric sections. The physical significance of this intuitive result is that if a circle is deformed in a highly eccentric elliptical section but the cross-sectional area is kept constant, then the major-axis will be much larger than the circle diameter which implies that it is easier to excite the higher-order even circumferential modes or they become cut-on at a lower frequency. Further note that the resonance or cut-on frequency of the $(1, 1)e$, i.e., the first higher-order mode, decreases significantly for a highly eccentric elliptical duct belonging to the iso-area family, and indeed, this

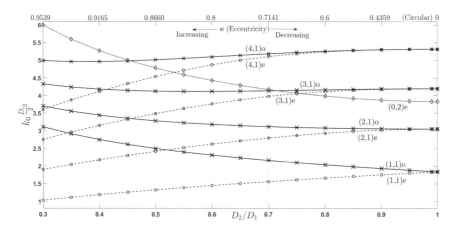

Fig. 2.11 Variation of non-dimensional resonance frequency of the first few even and odd circumferential modes and the first radial mode of a family of ellipses having equal cross-sectional area with D_2/D_1

result is even more pronounced for the $(4, 1)e$ mode graph. Therefore, from the point-of-view of a passive noise control device, it is rather unrewarding to use an iso-area family of elliptical ducts due to the early onset of higher-order even circumferential modes. Additionally, note that the graphs for lower-order odd circumferential modes and the first radial $(0, 2)e$ mode show a concave variation similar to Fig. 2.5. However, for higher-order odd circumferential modes, the concavity decreases significantly and the graphs are nearly constant, see for instance the $(3, 1)o$ and $(4, 1)o$ mode graphs wherein it is observed that the resonance frequencies of a circular and a highly eccentric same area elliptical duct are nearly the same! This rather remarkable result seems to suggest that even major deformations of the circular duct which retains the same cross-sectional area will have virtually no influence on the cut-off conditions of the sound radiated insofar as the contribution due to the higher-order odd circumferential modes is concerned [35].

A practical implication of the results analyzed above is that if the walls of the iso-area elliptical ducts are lined with a sound-absorbing material, the attenuation is likely to increase because the walls come closer. Secondly, there could be possible advantages to be gained from the directionality pattern of the sound radiated into the atmosphere from an elliptic duct, which could allow some azimuthal redistribution of the noise field [35]. This monograph, however, limits the investigation to the analysis of short elliptical chamber mufflers; therefore, the implications of the above results on the broadband attenuation (transmission loss) range of such muffler configurations will be briefly discussed in Chap. 4.

Finally, it may be mentioned that there is yet another family of ellipses which have the same semi-interfocal distance; for such a family, the non-dimensional resonance frequency expression pertaining to the even and odd modes is given by

$$(k_0 h)_{(m,n) \text{ Even}} = 2\sqrt{q_{m,n}} \qquad (2.76)$$

and

$$(k_0 h)_{(m,n)\text{odd}} = 2\sqrt{\overline{q}_{m,n}} \qquad (2.77)$$

A rather obvious corollary of Eqs. (2.76, 2.77) is that for a nearly circular duct (eccentricity tending to zero) belonging to this family, the higher-order modes are excited or get cut-on at much lower frequencies due to the large size of major-axis. This follows from the relation $h = (D_1/2)e$ and by recalling that two foci do not coalesce because the semi-interfocal distance h is kept constant. In this case, ξ tends to a very large value which is evident from the relation $\xi_0 = \cosh^{-1}(1/e)$.

To summarize the foregoing developments, this chapter has presented the analytical solution of the 3-D acoustic field in an infinite rigid-wall elliptical cylindrical waveguide in terms of angular and radial Mathieu functions. The tabulated transverse resonance frequencies corresponding to the radial, even and odd modes for a wide range of aspect-ratios D_2/D_1 provide an important set of data which can be used by students of engineering acoustics as well as muffler designers. Both analytical solution and resonance frequency data (Tables 2.8, 2.9, 2.10, 2.11, 2.12, 2.13, 2.14, 2.15, 2.16, 2.17, 2.18, 2.19, 2.20, 2.21, 2.22, 2.23, 2.24, 2.25, 2.26) are a prerequisite for 3-D analysis of elliptical mufflers. In fact, an analysis of the mode shapes revealed the location of pressure nodal hyperbolas (diameters) and ellipses (circles), the knowledge of which is important to determine the optimal location of inlet and outlet ports. Therefore, this chapter sets the foundation for the ensuing chapters.

Appendix: Resonance Frequencies of a Clamped Elliptical Membrane

The non-dimensional even and odd resonance frequencies of the vibration of a clamped elliptical membrane denoted by $\left(k_0 \dfrac{D_1}{2}\right)\bigg|_{(m,n)\ \text{Even}}$ and $\left(k_0 \dfrac{D_1}{2}\right)\bigg|_{(m,n)\ \text{odd}}$, respectively, were found by numerically solving Eqs. (2.62) and (2.63) for the parametric zeros $q_{m,n}$ and $\overline{q}_{mn}$, respectively, using the root-bracketing and bisection techniques explained in Sect. 2.5. Note that the parametric zeros computed here pertain to the Dirichlet boundary condition specified by Eqs. (2.62) and (2.63), and they are always interlaced with the parametric zeros corresponding to the Neumann or rigid-wall duct condition specified by Eqs. (2.49) and (2.50). The interlacing of parametric zeros is demonstrated by Fig. 2.3b which shows the variation of the functions $Ce_0(\xi = \xi_0, q)$ and $\frac{d}{d\xi}Ce_0(\xi = \xi_0, q)$ with the parameter q for $D_2/D_1 = 0.6$.

Table 2.29 presents the non-dimensional resonance frequencies of the radial mode $Ce_0(\xi, q_{0,n})ce_0(\eta, q_{0,n})$ and the even circumferential modes $Ce_m(\xi, q_{m,n})ce_m(\eta, q_{m,n})$ for the first seven orders, i.e., $m = 1, 2, \ldots 7$, while Table 2.30 presents the non-dimensional resonance frequencies of the first eight odd circumferential modes up to four decimal places. For each mode type, the first four zeros are presented, i.e., $n = 1, 2, 3, 4$. Note that a complete range of aspect-ratio given by

$$\frac{D_2}{D_1} = 0.1, 0.2, (0.1), \ldots, 0.8, 0.9, 1(\text{Circular}), \tag{2.78}$$

is considered in Tables 2.29 and 2.30, and the non-dimensionalization variable D_1 is kept constant in these Tables.

For certain aspect-ratio, the non-dimensional resonance frequencies shown in Tables 2.29 and 2.30 were compared with the values available in the literature for a clamped elliptical membrane [25, 26, 31] whereby an excellent agreement was observed. The graphical variation of the resonance frequencies presented in Tables 2.29 and 2.30 with aspect-ratio is not included for brevity; however, it may be mentioned that the graphs for both even and odd circumferential and cross-modes

Table 2.29 Non-dimensional resonance frequencies of *radial* and *even* circumferential/cross-modes of an elliptical membrane clamped at the boundaries for a range of aspect-ratio

Even Modes $Ce_m\left(\xi, q_{m,n}\right)\big\|_{\xi=\xi_0}=0$	$\dfrac{D_2}{D_1}$	$m=0$	$m=1$	$m=2$	$m=3$	$m=4$	$m=5$	$m=6$	$m=7$
$n=1$	0.1	16.2257	17.2604	18.3247	19.4171	20.5360	21.6798	22.8471	24.0365
	0.2	8.3936	9.4693	10.6003	11.7802	13.0030	14.2633	15.5564	16.8780
	0.3	5.8024	6.9266	8.1269	9.3889	10.7001	12.0506	13.4324	14.8394
	0.4	4.5257	5.7040	6.9708	8.3015	9.6780	11.0878	12.5217	13.9738
	0.5	3.7772	5.0102	6.3335	7.7142	9.1317	10.5730	12.0299	13.4966
	0.6	3.2933	4.5773	5.9440	7.3560	8.7924	10.2406	11.6921	13.1404
	0.7	2.9601	4.2892	5.6855	7.1096	8.5398	9.9611	11.3628	12.7394
	0.8	2.7202	4.0878	5.4979	6.9070	8.2881	9.6293	10.9356	12.2173
	0.9	2.5416	3.9415	5.3379	6.6792	7.9634	9.2120	10.4385	11.6491
	1.0	2.4048	3.8317	5.1356	6.3802	7.5883	8.7715	9.9361	11.0864
$n=2$	0.1	47.6301	48.6425	49.6653	50.6982	51.7412	52.7940	53.8564	54.9283
	0.2	24.0764	25.1048	26.1535	27.2216	28.3085	29.4133	30.5355	31.6744
	0.3	16.2326	17.2810	18.3588	19.4643	20.5957	21.7516	22.9302	24.1302
	0.4	12.3181	13.3907	14.5010	15.6458	16.8222	18.0271	19.2581	20.5127
	0.5	9.9771	11.0787	12.2252	13.4114	14.6329	15.8857	17.1662	18.4719
	0.6	8.4248	9.5615	10.7491	11.9811	13.2523	14.5588	15.8982	17.2681
	0.7	7.3260	8.5072	9.7469	11.0392	12.3801	13.7641	15.1818	16.6218
	0.8	6.5161	7.7623	9.0819	10.4640	11.8873	13.3273	14.7629	16.1775
	0.9	5.9153	7.2737	8.6982	10.1380	11.5523	12.9205	14.2489	15.5507
	1.0	5.5201	7.0156	8.4172	9.7610	11.0647	12.3386	13.5893	14.8213
$n=3$	0.1	79.0439	80.0520	81.0663	82.0869	83.1136	84.1464	85.1852	86.2300
	0.2	39.7798	40.7995	41.8315	42.8757	43.9316	44.9991	46.0780	47.1680
	0.3	26.6976	27.7327	28.7859	29.8568	30.9446	32.0488	33.1688	34.3040
	0.4	20.1626	21.2173	22.2958	23.3968	24.5193	25.6622	26.8243	28.0046
	0.5	16.2482	17.3277	18.4361	19.5715	20.7319	21.9157	23.1212	24.3469
	0.6	13.6463	14.7572	15.9016	17.0766	18.2797	19.5082	20.7601	22.0336
	0.7	11.7971	12.9489	14.1384	15.3616	16.6156	17.8977	19.2068	20.5423
	0.8	10.4234	11.6320	12.8829	14.1737	15.5045	16.8768	18.2877	19.7259
	0.9	9.3785	10.6823	12.0464	13.4663	14.9153	16.3596	17.7728	19.1478
	1.0	8.6537	10.1735	11.6198	13.0152	14.3725	15.7002	17.0038	18.2876
$n=4$	0.1	110.4589	111.4651	112.4759	113.4911	114.5107	115.5348	116.5632	117.5960
	0.2	55.4859	56.5018	57.5267	58.5603	59.6026	60.6534	61.7126	62.7799
	0.3	37.1667	38.1962	39.2388	40.2943	41.3623	42.4426	43.5349	44.6387
	0.4	28.0127	29.0599	30.1244	31.2057	32.3030	33.4159	34.5438	35.6861
	0.5	22.5265	23.5967	24.6881	25.7997	26.9306	28.0798	29.2465	30.4296
	0.6	18.8763	19.9762	21.1010	22.2492	23.4192	24.6098	25.8197	27.0476
	0.7	16.2783	17.4176	18.5851	19.7788	20.9967	22.2369	23.4980	24.7786
	0.8	14.3427	15.5374	16.7635	18.0186	19.3007	20.6089	21.9436	23.3071
	0.9	12.8602	14.1444	15.4666	16.8305	18.2404	19.6863	21.1430	22.5839
	1.0	11.7915	13.3237	14.7960	16.2235	17.6160	18.9801	20.3208	21.6415

Table 2.30 Non-dimensional resonance frequencies of *odd* circumferential/cross-modes of an elliptical membrane clamped at the boundaries for a range of aspect-ratio

Odd Modes $Se_m\left(\xi,\overline{q}_{m,n}\right)\Big|_{\xi=\xi_0}=0$ $\dfrac{D_2}{D_1}$	$m=1$	$m=2$	$m=3$	$m=4$	$m=5$	$m=6$	$m=7$	$m=8$
$n=1$ 0.1	31.9250	32.9428	33.9760	35.0242	36.0869	37.1638	38.2545	39.3586
0.2	16.2282	17.2679	18.3372	19.4343	20.5578	21.7060	22.8774	24.0706
0.3	11.0058	12.0715	13.1795	14.3260	15.5074	16.7201	17.9611	19.2275
0.4	8.4039	9.4996	10.6483	11.8427	13.0766	14.3439	15.6398	16.9598
0.5	6.8518	7.9809	9.1702	10.4078	11.6841	12.9909	14.3218	15.6714
0.6	5.8256	6.9907	8.2173	9.4888	10.7926	12.1189	13.4603	14.8114
0.7	5.1008	6.3025	7.5598	8.8515	10.1627	11.4831	12.8056	14.1254
0.8	4.5646	5.8023	7.0794	8.3715	9.6637	10.9480	12.2214	13.4836
0.9	4.1542	5.4262	6.7049	7.9690	9.2130	10.4386	11.6491	12.8473
1.0	3.8317	5.1356	6.3802	7.5883	8.7715	9.9361	11.0864	12.2251
$n=2$ 0.1	63.3367	64.3464	65.3639	66.3892	67.4220	68.4624	69.5102	70.5653
0.2	31.9275	32.9505	33.9887	35.0420	36.1097	37.1915	38.2871	39.3959
0.3	21.4642	22.5042	23.5667	24.6508	25.7554	26.8796	28.0226	29.1833
0.4	16.2391	17.3005	18.3910	19.5089	20.6523	21.8197	23.0093	24.2197
0.5	13.1111	14.1988	15.3217	16.4769	17.6614	18.8728	20.1086	21.3667
0.6	11.0336	12.1541	13.3151	14.5121	15.7412	16.9988	18.2817	19.5868
0.7	9.5593	10.7218	11.9290	13.1748	14.4533	15.7592	17.0870	18.4313
0.8	8.4666	9.6856	10.9509	12.2516	13.5769	14.9160	16.2599	17.6014
0.9	7.6379	8.9365	10.2662	11.6051	12.9368	14.2530	15.5516	16.8337
1.0	7.0156	8.4172	9.7610	11.0647	12.3386	13.5893	14.8213	16.0378
$n=3$ 0.1	94.7513	95.7583	96.7706	97.7880	98.8106	99.8383	100.8711	101.9089
0.2	47.6327	48.6502	49.6781	50.7161	51.7641	52.8219	53.8894	54.9663
0.3	31.9319	32.9637	34.0108	35.0727	36.1491	37.2395	38.3435	39.4606
0.4	24.0873	25.1376	26.2080	27.2976	28.4056	29.5312	30.6737	31.8323
0.5	19.3869	20.4610	21.5595	22.6812	23.8247	24.9887	26.1721	27.3737
0.6	16.2608	17.3653	18.4983	19.6579	20.8421	22.0491	23.2772	24.5250
0.7	14.0371	15.1816	16.3584	17.5647	18.7979	20.0557	21.3361	22.6371
0.8	12.3824	13.5828	14.8189	16.0871	17.3838	18.7051	20.0462	21.4016
0.9	11.1182	12.4070	13.7317	15.0818	16.4441	17.8060	19.1581	20.4959
1.0	10.1735	11.6198	13.0152	14.3725	15.7002	17.0038	18.2876	19.5545
$n=4$ 0.1	126.1665	127.1722	128.1818	129.1954	130.2128	131.2341	132.2592	133.2882
0.2	63.3393	64.3541	65.3767	66.4070	67.4450	68.49041	69.5433	70.6034
0.3	42.4018	43.4295	44.4688	45.5194	46.5812	47.6538	48.7370	49.8306
0.4	31.9385	32.9834	34.0435	35.1184	36.2076	37.3108	38.4273	39.5569
0.5	25.6666	26.7338	27.8199	28.9239	30.0451	31.1828	32.3363	33.5048
0.6	21.4925	22.5890	23.7075	24.8468	26.0059	27.1837	28.3790	29.5910
0.7	18.5201	19.6555	20.8160	22.0000	23.2058	24.4321	25.6775	26.9409
0.8	16.3038	17.4941	18.7125	19.9567	21.2250	22.5158	23.8274	25.1581
0.9	14.6029	15.8817	17.1916	18.5290	19.8887	21.2624	22.6404	24.0138
1.0	13.3237	14.7960	16.2235	17.6160	18.9801	20.3208	21.6415	22.9452

as well as radial modes have a concave nature. As anticipated, the concavity is more pronounced for the odd mode graphs as compared to their even mode counterparts. Here, we present a 9th degree least-squares interpolating polynomial using which one can accurately predict the non-dimensional resonance frequency values for any aspect-ratio within the range $D_2/D_1 = [0.3, 1]$. It is given by

$$
\left(k_0 \frac{D_1}{2} \right) \bigg|_{m,n} = b_0 + b_1\beta + b_2\beta^2 + b_3\beta^3 + b_4\beta^4 + b_5\beta^5
$$
$$
+ b_6\beta^6 + b_7\beta^7 + b_8\beta^8 + b_9\beta^9, \qquad (2.79)
$$

where

$$
\beta = \left(\frac{D_2}{D_1} - 0.55 \right) / 0.30277, \qquad (2.80)
$$

and the coefficients $\{b_0, b_1, b_2, \ldots b_9\}$ for the first few radial, even and odd modes are listed in Tables 2.31 and 2.32.

Figure 2.12a–c presents the mode shapes of the first three radial modes given by $Ce_0(\xi, q_{0,1})ce_0(\eta, q_{0,1})$, $Ce_0(\xi, q_{0,2})ce_0(\eta, q_{0,2})$ and $Ce_0(\xi, q_{0,3})ce_0(\eta, q_{0,3})$, respectively, for the aspect-ratio $D_2/D_1 = 0.5$. Note here that the fundamental (0, 1)e mode exhibits a gradual variation across the elliptical cross-section and is characterized by a maximum at the center and zero displacement along the elliptical boundaries—indeed, this feature is in contrast with the fundamental or plane wave mode of a rigid-wall elliptical waveguide for which the acoustic pressure field is uniform across the cross-section. Therefore, if the wall of an elliptical waveguide is not rigid, rather, if it is lined with a sound absorbent or dissipative material, then the plane wave mode does not exist. Rather, the fundamental mode exhibits variation across the cross -section and has a dissipative part which implies it cannot propagate without attenuation [42]. In fact, by imposing the boundary conditions given by Eqs. (2.62) and (2.63), one obtains the pressure-release modes of an elliptical duct [47], while in case of lined ducts, these modes are more popularly referred to as the soft-wall modes [48]. The (0, 2)e and (0, 3)e radial mode shapes exhibit one and two pressure nodal ellipses, respectively, as their respective eigenfunctions are a solution of the S-L problem. The set of figures in the 2nd column, i.e., Fig. 2.12d–f, presents the first three even circumferential mode shapes given by $Ce_1(\xi, q_{1,1})ce_1(\eta, q_{1,1})$, $Ce_2(\xi, q_{2,1})ce_2(\eta, q_{2,1})$, and $Ce_3(\xi, q_{3,1})ce_3(\eta, q_{3,1})$, respectively, while the set of figures in the 3rd column, i.e., Fig. 2.12g–i, presents the first three odd circumferential mode shapes $Se_1(\xi, \bar{q}_{1,1})se_1(\eta, \bar{q}_{1,1})$, $Se_2(\xi, \bar{q}_{2,1})se_2(\eta, \bar{q}_{2,1})$, and $Se_3(\xi, \bar{q}_{3,1})se_3(\eta, \bar{q}_{3,1})$, respectively. The even circumferential mode-shapes shown

Table 2.31 Coefficients of a 9th degree least-squares polynomial fit for interpolating the non-dimensional resonance frequency of the first radial mode and first few even and odd circumferential modes

Coefficients	$(0, 1)e$	$(1, 1)e$	$(1, 1)o$	$(2, 1)e$	$(2, 1)o$	$(3, 1)e$	$(3, 1)o$	$(4, 1)e$	$(4, 1)o$
b_0	3.5119	4.7710	6.2922	6.1178	7.4391	7.5164	8.6476	8.9459	9.9036
b_1	−1.4521	−1.2974	−3.0813	−1.1661	−2.9726	−1.0714	−2.8590	−1.0139	−2.7557
b_2	0.84466	0.82047	1.6823	0.75784	1.6895	0.67326	1.6672	0.57901	1.6147
b_3	−0.46833	−0.47822	−0.92184	−0.47900	−0.92911	−0.47743	−0.94518	−0.4825	−0.9656
b_4	0.43312	0.44103	0.86308	0.44395	0.86276	0.44012	0.86227	0.38452	0.86215
b_5	−0.23705	−0.23904	−0.47527	−0.24689	−0.47422	−0.25547	−0.47438	−0.29668	−0.47725
b_6	−0.20239	−0.20374	−0.40337	−0.20312	−0.40346	−0.2189	−0.40499	−0.19705	−0.4088
b_7	0.11123	0.11242	0.22204	0.11239	0.22186	0.10085	0.22141	0.12699	0.22143
b_8	0.16820	0.16824	0.33619	0.16634	0.33622	0.17245	0.33616	0.17153	0.3374
b_9	−0.09259	−0.092759	−0.18507	−0.093195	−0.18505	−0.08818	−0.18511	−0.09225	−0.18471

Table 2.32 Coefficients of a 9th degree least-squares polynomial fit for interpolating the non-dimensional resonance frequency of the second radial mode and first few cross-modes

Coefficients	$(0, 2)e$	$(1, 2)e$	$(1, 2)o$	$(2, 2)e$	$(2, 2)o$	$(3, 2)e$	$(3, 2)o$
b_0	9.1312	10.249	11.979	11.416	13.083	12.624	14.224
b_1	−4.6619	−4.5561	−6.2391	−4.4318	−6.1401	−4.2945	−6.0252
b_2	2.5248	2.5585	3.3613	2.5850	3.3953	2.6088	3.4253
b_3	−1.3746	−1.3606	−1.8346	−1.3422	−1.8228	−1.3033	−1.8115
b_4	1.2995	1.3021	1.7304	1.3397	1.7359	1.4313	1.7426
b_5	−0.71097	−0.70654	−0.94948	−0.67073	−0.9467	−0.62171	−0.94410
b_6	−0.60458	−0.58434	−0.80615	−0.59082	−0.80378	−0.67581	−0.80515
b_7	0.33410	0.34439	0.44457	0.32129	0.44507	0.24896	0.44198
b_8	0.50566	0.4985	0.67258	0.49323	0.67153	0.5099	0.66964
b_9	−0.27712	−0.28161	−0.37010	−0.28006	−0.37061	−0.26235	−0.37101

in Fig. 2.12d–f exhibit pressure nodal hyperbolas which divide the elliptical cross-section in regions of alternating phase, and they are found to be similar to their counterpart mode shapes shown in Figs. 2.6a–c Similarly, the odd circumferential mode shapes shown in Fig. 2.12(g–i) resemble their counterpart mode shapes presented in Figs. 2.6d–f, respectively, as may be observed from the pressure nodal hyperbola pattern. An apparent difference between the Dirichlet and Neumann mode shapes is that the former is characterized by zero value of the function field along the elliptical boundaries and a local maximum is formed at the center of the same phase region, while the latter mode shape is characterized by formation of a local maximum near the elliptical boundary.

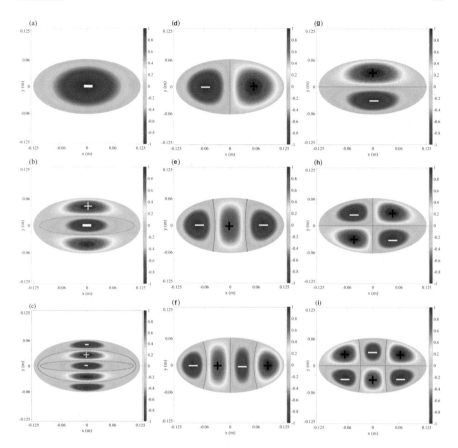

Fig. 2.12 First few mode shapes of a clamped vibrating elliptical membrane having $D_2/D_1 = 0.5$ Parts **a–c** show the first three radial (bouncing ball type) modes denoted by $(0, 1)e$, $(0, 2)e$ and $(0, 3)e$, respectively, parts **d–f** show the first three even circumferential (focus type) modes denoted by $(1, 1)e$, $(2, 1)e$ and $(3, 1)e$, respectively, while parts **g–i** show the first three odd circumferential (whispering gallery type) modes denoted by $(1, 1)o$, $(2, 1)o$ and $(3, 1)o$, respectively

References

1. E. Mathieu, Memoire sur le movement vibratoire d'une membrane de forme elliptique. J. Math. Pures Appl. **13**, 137 (1868)
2. E. Heine, *Handbuch der Kugelfunktionen*, 2 Vols. (1878/1881)
3. G. Floquet, Sur les equations differentielles lineaires. Ann. De l'Ecole Normale Superieure **12**, 47 (1883)
4. G.W. Hill, Mean equation of the lunar perigee. Acta. Math. **8**, 1 (1886)
5. N.W. McLachlan, *Theory and Application of Mathieu Functions* (Oxford University Press, London, 1947).
6. J. Meixner, F.W. Schäfke, *Mathieusche Funktionen Und Sphäroidfunktionen* (Springer-Verlag, Berlin, 1954).
7. M. Abramowitz, I.A. Stegun, *Handbook of Mathematical Functions with Formulas, Graphs, and Mathematical Tables* (National Bureau of Standards, Washington, DC, 1972).

8. E.L. Ince, Researches into the characteristic numbers of the mathieu equation. Proc. Roy. Soc. Edim. **46**, 20–29 (1926)
9. E.T. Kirkpatrick, Tables of values of the modified Mathieu functions. Math. Comput. **14**, 118–129 (1970)
10. S. Goldstein, On the asymptotic expansion of the characteristic number of the Mathieu equation. Proc. Roy. Soc. Edim. **49**, 203–223 (1929)
11. J. Canosa, Numerical solution of Mathieu's equation. J. Comput. Phys. **7**, 255–272 (1971)
12. D. Frenkel, R. Portugal, Algebraic methods to compute Mathieu functions. J. Phys. Math Gen **34**, 3541–3551 (2001)
13. F.A. Alhargan, A complete method for the computations of Mathieu characteristic numbers of integer orders. SIAM Rev. **38**, 239–255 (1996)
14. F.A. Alhargan, Algorithms for the computation of all Mathieu functions of integer orders. ACM Trans. Math. Software **26**, 390–407 (2000)
15. F.A. Alhargan, Algorithm 804: subroutines for the computation of Mathieu functions of integer orders. ACM Trans. Math. Softw. **26**, 408–414 (2000)
16. D. Clemm, Algorithm 352 Characteristic values and associated solutions of Mathieu's differential equation. Comm. ACM **12**, 399–408 (1969)
17. S.R. Rengarajan, J.E. Lewis, Mathieu functions of integral order and real arguments. IEEE Trans. Microw. Theory and Tech. MTT **28**, 276–277 (1980)
18. N. Toyama, K. Shogen, Computation of the value of the even and odd Mathieu functions of order n for a given parameter s and an argument x. IEEE T. Antenn. Propag. AP **32**, 537–539 (1994)
19. W. Leeb, Algorithm 537 Characteristic values of Mathieu's differential equation. ACM Trans. Math. Soft. **5**, 112–117 (1979)
20. R.B. Shirts, Algorithm 721 MTIEU1 and MTIEU2: Two subroutines to compute eigenvalues and solutions to Mathieu's differential equation for non-integer and integer order. ACM Trans. Math. Soft. **19**, 391–406 (1993)
21. J.J. Stamnes, B. Spjelkavik, New method for computing eigenfunctions (Mathieu functions) for scattering by elliptical cylinders. Pure Appl. Opt. **4**, 251–262 (1995)
22. S. Zhang, J. Jin, *Computation of Special Functions* (Wiley, New York, 1996).
23. C. Julio, *Gutiérrez-Vega, Formal Analysis of the Propagation of Invariant Optical Fields in Elliptic Coordinates*, Ph. D. Thesis, INAOE, México (20000
24. E. Cojocaru, *Mathieu Functions Computational Toolbox Implemented in MATLAB*. https:// arxiv.org/abs/0811.1970, https://www.mathworks.com/matlabcentral/fileexchange/22081-mat hieu-functions-toolbox-v-1-0
25. B.A. Troesch, H.R. Troesch, Eigenfrequencies of an elliptic membrane. Math. Comp. **27**, 755–765 (1973)
26. G. Chen, P.J. Morris, J. Zhou, Visualization of special eigenmode shapes of a vibrating elliptical membrane. SIAM Rev. **36**, 453–469 (1994)
27. J.B. Keller, S.I. Rubinow, Asymptotic solution of eigenvalue problems. Ann. Phys. **9**, 24–75 (1960)
28. J.C. Gutiérrez-Vega, R.M. Rodríguez-Dagnino, M.A. Meneses-Nava, S. Chávez-Cerda, Mathieu functions, a visual approach. Am. J. Phys. **71**, 233–242 (2003)
29. S. Ancey, A. Folacci, P. Gabrielli, Whispering-gallery modes and resonances of an elliptic cavity. J. Phys. A Math. Gen. **34**, 1341–1359 (2001)
30. L.D. Akulenko, S.V. Nesterov, Free vibrations of a homogeneous elliptic membrane. Mech. Solids **35**, 153–162 (2000)
31. H.B. Wilson, R.W. Scharstein, Computing elliptic membrane high frequencies by Mathieu and Galerkin methods. J. Eng. Math. **57**, 41–55 (2007)
32. L.J. Chu, Electromagnetic waves in elliptic hollow metal pipes of metal. J. Appl. Phys. **9**, 583–591 (1938)
33. S.D. Daymond, The principal frequencies of vibrating systems with elliptic boundaries. J. Mech. Appl. Math. **VIII**, 361–372 (1955)

34. J.G. Kretzschmar: Wave propagation in hollow conducting elliptical waveguides. IEEE Trans. Microw. Theory MTT **18** (1970)
35. M.V. Lowson, S. Baskaran, Propagation of sound in elliptic ducts. J. Sound Vib. **38**, 185–194 (1975)
36. F.D. Denia, J. Albelda, F.J. Fuenmayor, A.J. Torregrosa, Acoustic behaviour of elliptical chamber mufflers. J. Sound Vib. **241**, 401–421 (2001)
37. A. Mimani, M.L. Munjal, 3-D acoustic analysis of elliptical chamber mufflers having an end inlet and a side outlet: an impedance matrix approach. Wave Motion **49**, 271–295 (2012)
38. A. Mimani, M.L. Munjal, Acoustic end-correction in a flow-reversal end chamber muffler: a semi-analytical approach. J. Comput. Acoust. **24**, 1650004 (2016)
39. K. Hong, J. Kim, Natural mode analysis of hollow and annular elliptical cylindrical cavities. J. Sound Vib. **183**, 327–351 (1995)
40. W.M. Lee., Acoustic eigenproblems of elliptical cylindrical cavities with multiple elliptical cylinders by using the collocation multipole method. Int. J. Mech. Sci. **78**, 203–214 (2014)
41. M.L. Munjal, *Acoustics of Ducts and Mufflers* , 2nd edn. (Wiley, Chichester, UK, 2014).
42. D.T. Blackstock, *Fundamentals of Physical Acoustics*, Chap. 3 (John Wiley, New York, 2000)
43. J.M.G.S. Oliveira, P.J.S. Gil, Sound propagation in acoustically lined elliptical ducts. J. Sound Vib. **333**, 3743–3758 (2014)
44. M. Willatzen, L.C.L. Yan Voon, Flow-acoustic properties of elliptical-cylinder waveguides and enclosures. J. Phys. Conf. Ser. **52**, 1–13 (2006)
45. G.B. Arfken, H.J. Weber, *Mathematical Methods for Physicists* (Academic Press Elsevier, London, 2005), pp. 869–879
46. E. Kreyszig, *Advanced Engineering Mathematics*, 10th edn. (John Wiley & Sons, New Jersey, USA, 2011).
47. A. Sarkar, V.R. Sonti, Wave equations and solutions of in-vacuo and fluid-filled elliptical cylindrical shells. Int. J. Acoust. Vib. **14**, 35–45 (2009)
48. S.W. Rienstra, B.J. Tester, An analytic Green's function for a lined circular duct containing uniform mean flow. J. Sound Vib. **317**, 994–1016 (2008)

Chapter 3
Characterization of an Elliptical Chamber Muffler

The acoustic response function for an elliptical muffler having arbitrary port location is obtained by first finding the acoustic pressure field inside a rigid-wall elliptical cylindrical cavity as summation of orthogonal modes. To this end, we set $M_0 = 0$ in Eq. (2.46) implying a stationary medium, and on imposing zero normal velocity on the end faces $z = 0$ and L, one obtains

$$p(\xi, \eta, z, t) = \left\{ \begin{aligned} &\sum_{P=0,1,2,\ldots} \sum_{m=0,1,2\ldots}^{\infty} \sum_{n=1,2\ldots}^{\infty} C_{P,m,n} Ce_m(\xi, q_{m,n}) \times \\ &ce_m(\eta, q_{m,n}) \cos\left(\frac{P\pi z}{L}\right) \\ &+ \sum_{P=0,1,2,\ldots}^{\infty} \sum_{m=1,2\ldots}^{\infty} \sum_{n=1,2\ldots}^{\infty} S_{P,m,n} Se_m(\xi, \bar{q}_{m,n}) \times \\ &se_m(\eta, \bar{q}_{m,n}) \cos\left(\frac{P\pi z}{L}\right) \end{aligned} \right\} e^{j\omega t}, \qquad (3.1)$$

where $C_{P,m,n}$ and $S_{P,m,n}$ are the modal amplitudes [1]. Equation (3.1) is a general solution of the homogeneous Helmholtz equation (in elliptical cylindrical coordinates) subject to homogeneous rigid-wall boundary conditions on the end faces and side surface.

Assuming a zero mean flow for analyzing the acoustic attenuation performance of automotive mufflers is not uncommon: Typically, in the exhaust and tail pipes, the mean flow Mach number $M_0 \approx 0.15$, therefore, in the expansion chamber or muffler proper (Fig. 2.6 of Ref. [2]), the Mach number would be even less. Consequently, at such low Mach number, the convective effect of flow is negligible and thus can be safely ignored, see Refs. [2, 3].

A. Mimani, *Acoustic Analysis and Design of Short Elliptical End-Chamber Mufflers*, https://doi.org/10.1007/978-981-10-4828-9_3

3.1   Three-Dimensional Green's Function

The 3-D Green's function response for a rigid-wall elliptical cylindrical chamber is obtained by considering the inhomogeneous Helmholtz equation where the inlet (excitation) port is modeled by a point volume source coincident with the port center.

$$\left(\nabla^2 + k_0^2\right)p = -j\omega\rho_0 Q_0 \frac{\delta(\xi - \xi_0)}{h_\xi} \frac{\delta(\eta - \eta_0)}{h_\eta} \delta(z - z_0), \qquad (3.2)$$

where $\delta(\xi - \xi_0)/h_\xi$ and $\delta(\eta - \eta_0)/h_\eta$ denote Dirac delta functions in elliptical coordinates. Further, $Q_0 = U_0 S_{port}$ is the flow rate due to a port approximated as a point-source, while U_0 is the piston velocity (assumed uniform) and S_{port} is the port cross-sectional area. Equation (3.1) is now inserted in Eq. (3.2) to obtain

$$\sum_{P=0,1,2\ldots}^{\infty} \sum_{m=0,1,2\ldots}^{\infty} \sum_{n=1,2,\ldots}^{\infty} K_{P,m,n}^2 C_{P,m,n} Ce_m\left(\xi, q_{m,n}\right) ce_m\left(\eta, q_{m,n}\right) \cos\left(\frac{P\pi z}{L}\right)$$

$$+ \sum_{P=0,1,2,\ldots}^{\infty} \sum_{m=1,2,\ldots}^{\infty} \sum_{n=1,2,\ldots}^{\infty} \bar{K}_{P,m,n}^2 S_{P,m,n} Se_m\left(\xi, \bar{q}_{m,n}\right) se_m\left(\eta, \bar{q}_{m,n}\right) \cos\left(\frac{P\pi z}{L}\right)$$

$$= j\omega\rho_0 Q_0 \frac{\delta(\xi - \xi_0)}{h_\xi} \frac{\delta(\eta - \eta_0)}{h_\eta} \delta(z - z_0), \qquad (3.3)$$

where

$$K_{P,m,n} = \sqrt{\left(\frac{4q_{m,n}}{h^2} + \left(\frac{P\pi}{L}\right)^2 - k_0^2\right)}, \quad \bar{K}_{P,m,n} = \sqrt{\left(\frac{4\bar{q}_{m,n}}{h^2} + \left(\frac{P\pi}{L}\right)^2 - k_0^2\right)}.$$

$$(3.4, 3.5)$$

The modal coefficients $C_{P,m,n}$ and $S_{P,m,n}$ are evaluated by multiplying both the sides of Eq. (3.3) by

(1) $h_\xi h_\eta Ce_{m1}(\xi, q_{m_1,n_1}) ce_{m_1}(\eta, q_{m_1,n_1}) \cos(P_1\pi z/L)$ for even modes, and

(2) $h_\xi h_\eta Se_{m_1}(\xi, \bar{q}_{m_1,n_1}) se_{m_1}(\eta, \bar{q}_{m_1,n_1}) \cos(P_1\pi z/L)$ for odd modes,

and integrating over the elliptical cylindrical volume given by $\xi = 0\ldots\xi_0, \eta = 0\ldots 2\pi, z = 0\ldots L$. The mode orthogonality of the circular functions and the product of Mathieu and modified Mathieu functions [4] imply that out of the infinite number of mode terms on LHS of Eq. (3.3), only the modal term corresponding to $m = m_1, n = n_1$ and $P = P_1$ is nonzero which yields the corresponding modal coefficient $C_{P,m,n}$ or $S_{P,m,n}$. These are substituted back in Eq. (3.1) to obtain the acoustic pressure response or the Green's function due to a point-source located either on the end face or side surface [1, 5, 6].

$$p(\xi_r, \eta_r, z_r | \xi_\theta, \eta_\theta, z_\theta) = G(\xi_r, \eta_r, z_r | \xi_\theta, \eta_\theta, z_\theta) = G(R_r | R_\theta)$$

$$= \rho_0 Q(jk_0c_0) \left\{ \begin{array}{l} \displaystyle\sum_{P=0,1,2,\ldots} \sum_{m=0,1,2,\ldots} \sum_{n=1,2,\ldots} \frac{\psi(\xi_r, \eta_r, z_r) \times \psi(\xi_\theta, \eta_\theta, z_\theta)}{\left\{ \left(\frac{P\pi}{L}\right)^2 + \frac{4q_{m,n}}{h^2} - k_0^2 \right\} N_{m,n,p}} \\[2em] + \displaystyle\sum_{P=0,1,2,\ldots} \sum_{m=1,2,\ldots} \sum_{n=1,2,\ldots} \frac{\overline{\psi(\xi_r, \eta_r, z_r)} \times \overline{\psi(\xi_\theta, \eta_\theta, z_\theta)}}{\left\{ \left(\frac{P\pi}{L}\right)^2 + \frac{4\bar{q}_{m,n}}{h^2} - k_0^2 \right\} \overline{N}_{m,n,p}} \end{array} \right\},$$

$$(3.6)$$

where

$$\psi(\xi_r, \eta_r, z_r) = Ce_m(\xi_r, q_{m,n}) ce_m(\eta_r, q_{m,n}) \cos\left(\frac{P\pi z_r}{L}\right),$$

$$\psi(\xi_\theta, \eta_\theta, z_\theta) = Ce_m(\xi_\theta, q_{m,n}) ce_m(\eta_\theta, q_{m,n}) \cos\left(\frac{P\pi z_\theta}{L}\right), \qquad (3.7, 3.8)$$

and

$$\overline{\psi(\xi_r, \eta_r, z_r)} = Se_m(\xi_r, \bar{q}_{m,n}) se_m(\eta_r, \bar{q}_{m,n}) \cos\left(\frac{P\pi z_r}{L}\right),$$

$$\overline{\psi(\xi_\theta, \eta_\theta, z_\theta)} = Se_m(\xi_\theta, \bar{q}_{m,n}) se_m(\eta_\theta, \bar{q}_{m,n}) \cos\left(\frac{P\pi z_\theta}{L}\right). \qquad (3.9, 3.10)$$

while R_r and R_θ denote the receiver and source, respectively. Further, $N_{m,n,P}$ and $\overline{N}_{m,n,P}$ are the integral of the square of the product of a particular set of orthogonal mode functions defined over the elliptical cylindrical volume. Their analytical evaluation is tedious [4], and hence, these are computed numerically using the following integrals [1, 5]:

$$N_{m,n,P} = \int_{z=0}^{L} \int_{\xi=0}^{\xi_0} \int_{\eta=0}^{2\pi} h_\xi h_\eta Ce_m^2(\xi, q_{m,n}) ce_m^2(\eta, q_{m,n}) \cos^2\left(\frac{P\pi z}{L}\right) d\xi \, d\eta \, dz, \quad (3.11)$$

$$\overline{N}_{m,n,P} = \int_{z=0}^{L} \int_{\xi=0}^{\xi_0} \int_{\eta=0}^{2\pi} h_\xi h_\eta Se_m^2(\xi, \bar{q}_{m,n}) se_m^2(\eta, \bar{q}_{m,n}) \cos^2\left(\frac{P\pi z}{L}\right) d\xi \, d\eta \, dz. \quad (3.12)$$

Nevertheless, a proof of the mode orthogonality of the product of Mathieu and modified Mathieu functions is presented in Section 9.4 of Mclachlan [4]: The integral $\int_0^{\xi_0} \int_0^{2\pi} h_\xi h_\eta \zeta_{m,n} \zeta_{p,r} d\xi \, d\eta$ is nonzero only if $m = p, n = r$ where $\zeta_{m,n} = Ce_{m,n} ce_{m,n}$ or $Se_{m,n} se_{m,n}$ and $\zeta_{p,r} = Ce_{p,r} ce_{p,r}$ or $Se_{p,r} se_{p,r}$. In Eq. (3.6),

L is the chamber length, while the source and receiver coordinates are denoted by $(\xi_\theta, \eta_\theta, z_\theta)$ and (ξ_r, η_r, z_r), respectively.

In the limit of the elliptical cross-section approaching a circular one, the even and odd modes coalesce implying $Ce_m(\cdot) \to Se_m(\cdot) \to J_m(\cdot)$, $ce_m(\cdot) \to \cos(m\theta)$ and $se_m(\cdot) \to \sin(m\theta)$. Consequently, the 3-D Green's function for a circular cylindrical chamber is given by [7–9]

$$\frac{p(r_R,\theta_R,z_R|r_S,\theta_S,z_S)}{\rho_0 Q_0} = \frac{G(r_R,\theta_R,z_R|r_S,\theta_S,z_S)}{\rho_0 Q_0} =$$

$$jk_0c_0 \left\{ \sum_{P=0,1,2,\ldots}^{\infty} \sum_{m=0,1,2,\ldots}^{\infty} \sum_{n=0,1,2,\ldots}^{\infty} \frac{J_m\left(\alpha_{mn}\frac{r_R}{R_0}\right)\cos\left(\frac{P\pi z_R}{L}\right)J_m\left(\alpha_{mn}\frac{r_S}{R_0}\right)\cos\left(\frac{P\pi z_S}{L}\right)\cos(m(\theta_R-\theta_S))}{\left\{\left(\frac{P\pi}{L}\right)^2+\left(\frac{\alpha_{mn}}{R_0}\right)^2-k_0^2\right\}N_{l,m,n}} \right\}$$

$$(3.13)$$

where

$$N_{m,n,P}$$

$$= \left\{\int_{r=0}^{r=R_0} r\left(J_m\left(\frac{\alpha_{mn}}{R_0}r\right)\right)^2 dr\right\}\left\{\int_{\theta=0}^{\theta=2\pi}(\cos(m\theta))^2 d\theta\right\}\left\{\int_{z=0}^{z=L}\left(\cos\left(\frac{P\pi z}{L}\right)\right)^2 dz\right\},$$

$$(3.14)$$

is the integral of the square of the product of a particular set of mode shape functions defined over the circular cylindrical volume [7]. The integral $N_{m,n,P}$ is evaluated analytically; a closed-form expression for the first integral in Eq. (3.14) is given by $0.5R_0^2\{1-(m/\alpha_{mn})^2\}(J_m(\alpha_{mn}))^2$. Here, (r_R,θ_R,z_R) and (r_S,θ_S,z_S) denote the coordinates of the receiver port R and source port S, respectively, wherein the angular coordinates are measured with respect to the x-axis. Further, $\alpha_{m,n}$ is the non-dimensional resonance frequency of the (m, n) mode of the circular chamber, see Table 2.26.

It is observed from Eqs. (3.6) to (3.13) that on interchanging the location of source and receiver ports, the Green's function remains unaltered, thereby indicating that the 3-D acoustic field inside the chamber satisfies acoustic reciprocity [10]. Further, the Green's function response is purely imaginary showing that the empty rigid-wall elliptical (circular) chamber muffler (no dissipative lining) having zero mean flow is a reciprocal and conservative system [11].

3.2 Acoustic Pressure Response Based upon the Uniform Piston-Driven Model

Owing to its relative simplicity, the point-source or the Green's function response had been used in previous works to obtain the four-pole parameters of a three-dimensional annular cylindrical cavity [12], thin or two-dimensional mufflers of circular and rectangular cross-section [13] and an elliptical cylindrical muffler [6]. However, one of its main criticisms is the inherent modeling assumption; the inlet and outlet ports which are actually of a finite cross-section are considered as points on the cavity surface. The acoustic pressure response is thus evaluated at these points, thereby ignoring the fundamental conditions of continuity of normal acoustic velocity and acoustic pressure fields at the port-chamber interface [14, 15]. Furthermore, Zhou and Kim [16] demonstrated through the example of a rectangular cavity that the point-source model is not always valid and has singularity issues when the response and source points are coincident. In such cases, the auto-response function (same as self-impedance) does not converge even when a large number of modal terms are considered. The present author also experienced some convergence problems in the TL graph at high frequencies for a short chamber muffler when the Green's function was directly used.

With a view to improve the port-chamber interface modeling, a more accurate and realistic modeling approach: *the uniform piston-driven model is considered in this monograph* for characterizing elliptical and circular mufflers [1, 5, 7]. This method significantly improves upon the point-source model inasmuch as the actual (finite) cross-sectional area of the port is now considered. Here, the complete 3-D acoustic field is considered in the muffler cavity; however, the inlet/outlet ports are taken to be rigid oscillating pistons with uniform velocity distribution which is the same as considering only planar wave propagation in the ports, immediately from the port-chamber interface. The uniform piston velocity is taken equal to the normal acoustic velocity in the chamber over the port-chamber interface, whereas over the annular area, the normal acoustic velocity in the chamber is set to zero, thereby implying that this modeling approach satisfies the condition of continuity of velocity fields. The uniform piston model, therefore, ignores higher-order transverse modes in the ports or equivalently, near-field in the ports as its higher-order modes are generally evanescent for the frequency range of interest. In other words, this modeling approach does not consider modal coupling between the chamber and ports; the amplitudes of the chamber modes are explicitly evaluated in terms of the piston velocity U_0 through the velocity continuity condition and orthogonality relations.

While the uniform piston-driven model is accurate (at least for automotive muffler design) and indeed has been used in several papers [1, 5, 7, 17–23], by far the most accurate technique of modeling the acoustics of port-chamber interface is the analytical/numerical mode-matching (AMM or NMM) method [14, 15, 24–30]. This is because the AMM/NMM methods enforce both, the continuity of normal acoustic velocity as well as acoustic pressure fields across the port-chamber interface, thereby taking the modal coupling into account. The NMM method, in particular, can handle muffler chamber having a complicated internal structure where the 2-D transverse modes cannot be found analytically, for example, silencers with concentric straight-through perforated pipe (airway) whose annular volume is filled with dissipative material, see Refs. [27–30]. Both the AMM and NMM methods generally employ an anechoic termination at the outlet, and based on a point-collocation scheme [27] or using analytical modal orthogonality relations [24–26], they result in a system of algebraic equations for solving the modal coefficients of the chamber and ports in terms of the incident wave amplitude at the inlet. Based on the number of modes considered, often a large system of equations is obtained, and the associated algebraic manipulations can be tedious: the cost of achieving highest possible accuracy.

The uniform piston model can nevertheless still be used in favor of the AMM or NMM methods as it is simpler yet sufficiently accurate for analyzing the attenuation performance of most automotive mufflers. This is mainly because the port diameter d_0 is significantly smaller than diameter of the muffler shell (chamber) and it typically varies from 35 to 60 mm. In this monograph, a default value of $d_0 = 40$ mm is considered, and for sound speed $c_0 = 343.14 \, \mathrm{m \, s^{-1}}$, the first higher-order (1, 0) mode is cut-on at a high frequency $f = \dfrac{1.8412 c_0}{\pi d_0} \approx 5025$ Hz which is significantly greater than the frequency range of interest. This justifies the use of the uniform piston-driven model whose results are corroborated through a comparison with a full 3-D FEA of the chamber-port continuum, for specific muffler configurations [2].

We present two different ways to implement the piston-driven model for obtaining the pressure response function: First, an inhomogeneous Helmholtz equation is considered which has a source term with a uniform strength over the port area (modeling a piston) whereby the 3-D Green's function is integrated over the port to yield the response function. The second technique consists of solving the homogeneous Helmholtz equation subject to inhomogeneous boundary condition over the port-chamber interface.

3.2.1 Excitation at an End Port

The acoustic pressure response due to a uniform piston excitation at a port located on end face is first obtained, subsequent to which the average response over the receiver port will be computed. To this end, the inhomogeneous Helmholtz equation is considered [1],

$$\left(\nabla^2 + k_0^2\right)p = -j\omega\rho_0 U_{\mathrm{piston}} f_1(\xi, \eta)\delta(z - l_E), \tag{3.15}$$

where

$$f_1(\xi, \eta) = 1, \quad \forall S_{\mathrm{end}}$$
$$0. \quad \forall S_{\mathrm{chamber}} - S_{\mathrm{end}} \tag{3.16}$$

In Eq. (3.15), the suffix E denotes the end port, and in Eq. (3.16), S_{chamber} and S_{end} denotes the area of the end face of the elliptical chamber and the end port, respectively. Now, since Eq. (3.6) is a solution of Eq. (3.2), one obtains

$$\left(\nabla^2 + k_0^2\right)G(\xi, \eta, l|\xi_E, \eta_E, l_E) = -j\omega\rho_0 Q_0 \frac{\delta(\xi - \xi_E)}{h_\xi}\frac{\delta(\eta - \eta_E)}{h_\eta}\delta(z - l_E),$$

(3.17)

where ξ_E, η_E, l_E denote the coordinates pertaining to the center of the end port and we choose $l_E = 0$. Multiplying both sides of Eq. (3.17) with $h_\xi h_\eta f_1(\xi_E, \eta_E)$ and integrating over the elliptical cross-section (on which the end port is located), we get

$$\iint\limits_{S_{\text{chamber}}} F\{G(\xi, \eta, z|\xi_E, \eta_E, l_E)\}h_\xi h_\eta f_1(\xi_E, \eta_E)d\xi_E d\eta_E = -j\omega\rho_0 Q f_1(\eta, z)\delta(z - l_E),$$

(3.18)

where the operator $F = \left(\nabla^2 + k_0^2\right)$. On using the fact that the volume flow rate $Q_0 = U_0 S_{\text{end}}$, changing the order of the integration over the end port (piston) and operator F, and comparing the equation so obtained with Eq. (3.15), one realizes that uniform piston-driven pressure response due to the end port excitation is given by

$$p(\xi_r, \eta_r, z_r|\xi_E, \eta_E, l_E) = \frac{1}{S_{\text{end}}}\iint\limits_{S_{\text{end}}} G(\xi_r, \eta_r, z_r|\xi_E, \eta_E, l_E)h_\xi(\xi_E, \eta_E)h_\eta(\xi_E, \eta_E)d\xi_E d\eta_E.$$

(3.19)

3.2.1.1   Response at an End Port

The uniform piston-driven model enforces that over the excitation port area, the axial velocity profile u_z in the chamber is constant and equal to the acoustic velocity u_{end} in the port (at the interface), see Sect. 3.2.2.1. The acoustic pressure in the chamber over the excitation port cross-section (as well as the receiver port area), however, is not necessarily a constant and in fact may vary significantly depending on the chamber length L and frequency of excitation f_0, see Sect. 3.2.2.2. While Eq. (3.19) inherently takes care of this variation over the excitation port, we also need to account for the variation over the receiver port. Hence, we compute the average response by integrating the acoustic pressure response over the receiver port and dividing by its cross-sectional area. Here, we consider the case of a receiver end port to obtain the average pressure response as

$$p(\xi_{E2}, \eta_{E2}, l_{E2}|\xi_{E1}, \eta_{E1}, l_{E1})$$

$$= \frac{1}{S_{\text{end2}}} \iint\limits_{S_{\text{end2}}} \frac{1}{S_{\text{end1}}} \iint\limits_{S_{\text{end1}}} \left\{ G(\xi_{E2}, \eta_{E2}, l_{E2}|\xi_{E1}, \eta_{E1}, l_{E1}) \, h_\xi h_\eta \mathrm{d}\xi_{E1} \mathrm{d}\eta_{E1} \right\} h_\xi h_\eta \mathrm{d}\xi_{E2} \mathrm{d}\eta_{E2}.$$

$$(3.20)$$

In Eq. (3.20), if the end ports 2 and 1 are different, one would obtain the *cross-impedance* parameter, and when receiver/response port 2 is the same as the excitation port 1, we obtain the *self-impedance* parameter.

3.2.1.2   Response at a Side Port

For the case of receiver port located on the side surface of the elliptical chamber, one obtains the following cross-impedance parameter relating the end port excitation and the side port response.

$$p(\xi_S, \eta_S, l_S|\xi_E, \eta_E, l_E)$$

$$= \frac{1}{S_{\text{side}}} \iint\limits_{S_{\text{side}}} \frac{1}{S_{\text{end}}} \iint\limits_{S_{\text{end}}} \left\{ G(\xi_S, \eta_S, l_S|\xi_E, \eta_E, l_E) h_\xi h_\eta \mathrm{d}\xi_E \mathrm{d}\eta_E \right\} h_\eta \mathrm{d}l_S \mathrm{d}\eta_S, \quad (3.21)$$

where the coordinates ξ_S, η_S, l_S pertain to the location of the side port. Further, $\xi_0 = \xi_S$ is constant over the side surface, whereas l_E is constant over the end face. Due to reciprocity [10], Eq. (3.21) also holds good for the side port excitation and the end port response.

3.2.2   *Homogeneous Wave Equation Subject to Inhomogeneous Conditions*

The acoustic pressure response functions due to piston excitation and response ports located at the end faces of a rigid-wall elliptical cylindrical chamber may alternatively be derived by solving the homogenous Helmholtz equation subjected to the following inhomogeneous boundary conditions on the end face [1].

$$u_z = \frac{-1}{jk_0 \rho_0 c_0} \frac{\mathrm{d}}{\mathrm{d}z} p(\xi, \eta, z)\bigg|_{z=0} = 0, \quad \forall S_{\text{chamber}} - S_{\text{end}}$$

$$= u_{\text{end}} = U_0 \quad \forall S_{\text{end}}. \quad (3.22)$$

$$p(\xi, \eta, z = 0) = p_{\text{end}}. \quad \forall S_{\text{end}} \quad (3.23)$$

Additionally, for the opposite end face, we have the homogenous boundary condition

$$u_z = \frac{-1}{jk_0\rho_0c_0}\frac{d}{dz}p(\xi,\eta,z)\bigg|_{z=L} = 0 \quad \forall S_{\text{chamber}}. \tag{3.24}$$

On making use of Eq. (3.24), we obtain the acoustic particle velocity u_z in the elliptical cylindrical chamber as

$$u_z = \left(\frac{1}{k_0\rho_0c_0}\right)\left\{\begin{array}{l}\displaystyle\sum_{m=0,1,2,\ldots}^{\infty}\sum_{n=1,2,\ldots}^{\infty}C_{m,n}^1 k_{z,m,n}Ce_m(\xi,q_{m,n})ce_m(\eta,q_{m,n})\times \\[2mm] \left(e^{-jk_{z,m,n}z} - e^{jk_{z,m,n}z}e^{-2jk_{z,m,n}L}\right) \\[3mm] \displaystyle+\sum_{m=1,2,\ldots}^{\infty}\sum_{n=1,2,\ldots}^{\infty}S_{m,n}^1 \bar{k}_{z,m,n}Se_m(\xi,\bar{q}_{m,n})se_m(\eta,\bar{q}_{m,n})\times \\[2mm] \left(e^{-j\bar{k}_{z,m,n}z} - e^{j\bar{k}_{z,m,n}z}e^{-2j\bar{k}_{z,m,n}L}\right)\end{array}\right\}, \tag{3.25}$$

where $k_{z,m,n}$ and $\bar{k}_{z,m,n}$ are obtained by setting $M_0 = 0$ in Eqs. (2.54) and (2.55), respectively. On substituting Eq. (3.25) into Eqs. (3.22) and (3.23) and then, making use of orthogonality of Mathieu functions [4], we obtain the modal coefficients $C_{m,n}^1$ and $S_{m,n}^1$ which are substituted back in the acoustic pressure field expression inside the elliptical cylindrical chamber, i.e., in Eq. (2.53) (setting $M_0 = 0$) to yield

$$p_{\text{chamber}}(\xi,\eta,z)$$

$$=\left\{\begin{array}{l}\displaystyle\sum_{m=0,1,2\ldots}^{\infty}\sum_{n=1,2\ldots}^{\infty}C_{m,n}^1 Ce_m(\xi,q_{m,n})ce_m(\eta,q_{m,n})\left(e^{-jk_{z,m,n}z} + e^{jk_{z,m,n}z}e^{-2jk_{z,m,n}L}\right) \\[3mm] \displaystyle+\sum_{m=1,2\ldots}^{\infty}\sum_{n=1,2\ldots}^{\infty}S_{m,n}^1 Se_m(\xi,\bar{q}_{m,n})se_m(\eta,\bar{q}_{m,n})\left(e^{-j\bar{k}_{z,m,n}z} + e^{j\bar{k}_{z,m,n}z}e^{-2j\bar{k}_{z,m,n}L}\right)\end{array}\right\}, \tag{3.26}$$

$$C_{m,n}^1$$

$$= U_0\left(\frac{k_0\rho_0c_0}{k_{z,m,n}}\right)\left\{\frac{\iint_{S_{\text{end1}}}Ce(\xi,q_{m,n})ce(\eta,q_{m,n})h_\xi h_\eta d\xi d\eta}{\left(1-e^{-2jk_{z,m,n}L}\right)\int_{\eta=0}^{2\pi}\int_{\xi=0}^{\xi=\xi_0}Ce^2(\xi,q_{m,n})ce^2(\eta,q_{m,n})h_\xi h_\eta d\xi d\eta}\right\}, \tag{3.27a}$$

and

$S_{m,n}^1$

$$= U_0 \left(\frac{k_0 \rho_0 c_0}{k_{z,m,n}} \right) \left\{ \frac{\iint_{S_{end1}} Se(\xi, \bar{q}_{m,n}) se(\eta, \bar{q}_{m,n}) h_\xi h_\eta d\xi d\eta}{\left(1 - e^{-2j\bar{k}_{z,m,n}L} \right) \int_{\eta=0}^{2\pi} \int_{\xi=0}^{\xi=\xi_0} Se^2(\xi, \bar{q}_{m,n}) se^2(\eta, \bar{q}_{m,n}) h_\xi h_\eta d\xi d\eta} \right\}.$$

$$(3.27b)$$

Equation (3.26) is now used to obtain the acoustic pressure response function (or the impedance [**Z**] matrix parameter) for the case of the end port excitation and the end port response.

$$p(\xi_{E2}, \eta_{E2}, z_{E2} | \xi_{E1}, \eta_{E1}, z_{E1}) = \frac{1}{S_{end2}} \iint_{S_{end2}} p_{chamber}(\xi_{E2}, \eta_{E2}, z_{E2}) h_\xi h_\eta d\xi d\eta.$$

$$(3.28)$$

It is noted that Eqs. (3.26) and (3.28) are indeed mathematically equivalent to Eqs. (3.19) and (3.20), respectively, as may be shown by using the trigonometric identity:

$$\frac{\cot \theta}{\theta} = \frac{1}{\theta^2} + \sum_{P=1,2,3,\ldots}^{\infty} \frac{2}{\theta^2 - (P\pi)^2}, \qquad (3.29)$$

where $\theta = k_{z,m,n}L$ or $\theta = \bar{k}_{z,m,n}L$, see Ref. [7]. Myint-U and Debnath [31] also show that solving an inhomogeneous Helmholtz equation subject to homogenous boundary condition is the same as solving an homogeneous Helmholtz equation subject to inhomogeneous boundary conditions. From the implementation point-of-view, the only difference between the Green's function-based piston model approach, i.e., Eqs. (3.19) and (3.26), is that in the former method, axial propagation is considered as discrete modes given by $\cos(P\pi z/L)$, $P = 0, 1, 2, \ldots$. Whereas in the latter, the axial propagation is modeled in terms of complex exponentials involving continuously varying wave numbers $k_{z,m,n}$ and $\bar{k}_{z,m,n}$. The latter approach is thus particularly suited for analyzing the end-inlet and end-outlet muffler configurations as can be appreciated from the end-correction analysis for long flow-reversal mufflers [7]. However, this approach is less suited for dealing with elliptical muffler configurations having a side port excitation because of the occurrence of a negative q parameter (in the Mathieu functions) corresponding to a given axial mode and excitation frequency, thereby rendering the resulting analysis, unnecessarily tedious. Similar remarks hold for a circular chamber with a side port excitation. Nevertheless, this method is useful in examining the modeling efficiency of the uniform piston approximation, i.e., acoustic velocity field u_z analysis in the chamber and its implications on evaluating the acoustic response function, i.e., the acoustic pressure profile analysis at the port-chamber interface.

3.2.2.1 Acoustic Velocity (Axial) Profile in Chamber at the Port-Chamber Interface

The modal coefficients $C_{m,n}^1$ and $S_{m,n}^1$ obtained in Eqs. (3.27a, b) are back substituted in Eq. (3.25) to obtain the ratio of the acoustic axial particle velocity $u_z(\xi, \eta, z = 0)$ to piston velocity $u_{end} = U_0$ at the end port shown hereunder.

$$
\frac{u_z}{u_{end}}
$$

$$
= \left\{
\begin{aligned}
&\sum_{m=0,1,2...}^{\infty} \sum_{n=1,2...}^{\infty} \frac{Ce_m(\xi, q_{m,n})ce_m(\eta, q_{m,n})\left(\iint_{Send} Ce(\xi, q_{m,n})ce(\eta, q_{m,n})h_\xi h_\eta d\xi d\eta\right)}{\int_{\eta=0}^{2\pi}\int_{\xi=0}^{\xi=\xi_0} Ce^2(\xi, q_{m,n})ce^2(\eta, q_{m,n})h_\xi h_\eta d\xi d\eta} \times \\
&\qquad\qquad \left(\frac{e^{-jk_{z,m,n}} - e^{jk_{z,m,n}(z-2L)}}{1 - e^{-2jk_{z,m,n}L}}\right) \\
&+\sum_{m=1,2...}^{\infty} \sum_{n=1,2...}^{\infty} \frac{Se_m(\xi, \bar{q}_{m,n})se_m(\eta, \bar{q}_{m,n})\left(\iint_{Send} Se(\xi, \bar{q}_{m,n})se(\eta, \bar{q}_{m,n})h_\xi h_\eta d\xi d\eta\right)}{\int_{\eta=0}^{2\pi}\int_{\xi=0}^{\xi=\xi_0} Se^2(\xi, \bar{q}_{m,n})se^2(\eta, \bar{q}_{m,n})h_\xi h_\eta d\xi d\eta} \times \\
&\qquad\qquad \left(\frac{e^{-j\bar{k}_{z,m,n}z} - e^{j\bar{k}_{z,m,n}(z-2L)}}{1 - e^{-2j\bar{k}_{z,m,n}L}}\right)
\end{aligned}
\right.
$$

$$(3.30)$$

It can be readily noted from Eq. (3.30) that $u_z(z = 0)/u_{end}$, i.e., at the port-chamber interface is independent of the excitation frequency or the wave numbers $k_{z,m,n}$ or $\bar{k}_{z,m,n}$ as well as the chamber length L.

Figure 3.1 shows the variation of $|u_z(z = 0)/u_{end}|$ ratio over the elliptical cross-section at the axial plane $z = 0$ due to a concentric end port obtained by considering the first nine orders of the even-even mode functions and their first nine parametric zeros, i.e., $Ce_{2m}(\xi, q_{m,n})ce_{2m}(\eta, q_{2m,n})$, $m = 0, 1, 2, ..., 8$ and $n = 1, 2, ..., 9$. This result demonstrates the efficiency of the uniform piston modeling; it is observed that while the $|u_z(z = 0)/u_E| \approx 1$ over the port cross-section, this ratio rapidly falls to zero over the annular region. This signifies that axial acoustic velocity in the chamber is equal to the piston velocity, roughly over most of the port area and then rapidly drops to zero at the rigid annular plate as is required by the conditions of continuity of particle velocity (Eq. 3.22). However, at some regions (over the port), $|u_z/u_{end}| > 1$ while at regions near the port boundaries, this ratio is about 0.75. These small fluctuations are indeed the well-known Gibbs phenomenon [32]; they occur despite considering large number of modes because the uniform piston model inherently attempts to mimic a discontinuous velocity profile in the chamber through a sequence of continuous functions.

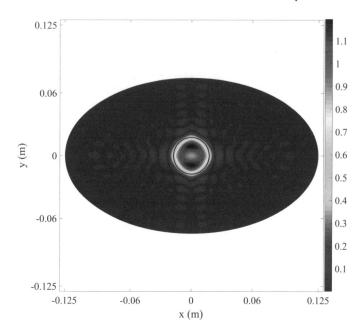

Fig. 3.1 Variation of $|u_z(\xi, \eta, z = 0)/u_{end}|$ ratio over the elliptical cross-section at axial plane $z = 0$ corresponding to the port-chamber interface. The elliptical cylindrical chamber has $D_1 = 0.25$ m, $D_2/D_1 = 0.6$, while the end port is concentric and its diameter $d_0 = 0.04$ m

3.2.2.2 Acoustic Pressure Field Response in Chamber at the Port-Chamber Interface

The non-dimensional impedance $p(\xi, \eta, z = 0)/\rho_0 c_0 U_0$ in the chamber at the port-chamber interface can be computed by virtue of Eqs. (3.26) and (3.27a, b). Unlike $|u_z(z = 0)/u_{end}|$ ratio, it is readily observed that $p(\xi, \eta, z = 0)/\rho_0 c_0 U_0$ is dependent on both the frequency of excitation f_0 and the chamber length L. Figure 3.2a, b show the normalized $|p(\xi, \eta, z = 0)/\rho_0 c_0 U_0|$ ratio over the elliptical cross-section for a long chamber with $L = 300$ mm at Helmholtz number $0.5k_0 D_1 = 1$(low frequency) and 7 (high frequency), respectively, while parts (c) and (d) show this variation for a short chamber $L = 50$ mm at $0.5k_0 D_1 = 1$ and 7, respectively. These non-dimensional impedance results demonstrate that while acoustic pressure field exhibits a small variation over the port area for long chambers, there is a significant variation over the port area for the case of short chambers, especially at high frequencies due to the influence of higher-order transverse modes that do not decay sufficiently [33, 34]. In general, the acoustic pressure field varies significantly over the elliptical cross-section at the $z = 0$ plane. Therefore, to accurately account for this variation, it is necessary to integrate the piston response (Eq. 3.19) over the port area (Eq. 3.20) to obtain the average pressure response which yields the impedance [**Z**] matrix parameter.

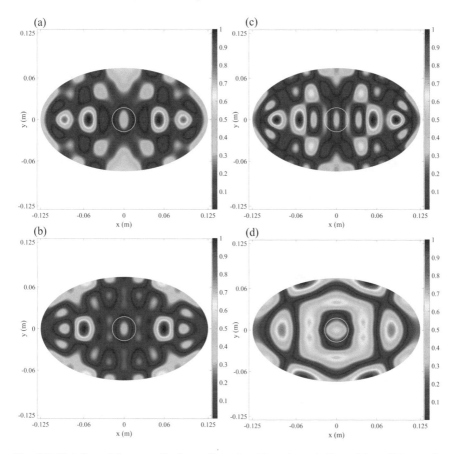

Fig. 3.2 Variation of the normalized non-dimensional impedance $|p(\xi, \eta, z)/\rho_0 c_0 U_0|$ over the elliptical cross-section at the port-chamber interface for (**a**, **b**) long chamber $L = 0.3$ m and (**c**, **d**) short chamber $L = 0.05$ m. The concentric end port configuration (Fig. 3.1) is considered here. Parts a and c are evaluated at $0.5 k_0 D_1 = 1$, while parts b and d are evaluated at $0.5 k_0 D_1 = 7$. The sound speed $c_0 = 343.14$ m · s^{-1} is considered for computing the results shown in (**a–d**)

3.2.3 Excitation at a Side Port

The inhomogeneous 3-D Helmholtz equation subject to a uniform velocity *elliptical piston* excitation at the side surface is now considered. (It is noted that the projection of the circular port on the side surface of an elliptical or circular cylindrical chamber has an elliptical cross-section, hence the term elliptic piston.)

$$\left(\nabla^2 + k_0^2\right)p = -j\omega\rho_0 U_{\text{piston}} f_2(\eta, z)\frac{\delta(\xi - \xi_S)}{h_\xi}, \qquad (3.31)$$

where

$$f_2(\eta, z) = 1, \quad \forall S_{\text{side}}$$
$$0 \quad \forall S_{\text{side surface}} - S_{\text{side}}, \tag{3.32}$$

and $S_{\text{side surface}}$ denotes the curved surface area of the elliptical (circular) chamber, while S_{side} denotes the elliptical piston area of the side port.

As with the end port excitation, it can also be shown that acoustic pressure response based upon the piston-driven model due to the side port excitation is given by

$$p(\xi_r, \eta_r, z_r | \xi_S = \xi_0, \eta_S, l_S)$$
$$= \frac{1}{S_{\text{side}}} \iint_{S_{\text{side}}} \{G(\xi_r, \eta_r, z_r | \xi_S = \xi_0, \eta_S, l_S)\} h_\eta(\xi_S, \eta_S) \mathrm{d}\eta_S \mathrm{d}l_S. \tag{3.33}$$

3.2.3.1 Response at a Side Port

Equation (3.33) is now integrated over the elliptical piston area of the receiver side port (which is not necessarily the same as the excitation side port in Eq. (3.31)) and divided by its cross-sectional area to yield the average acoustic pressure response given by

$$p(\xi_r = \xi_0, \eta_{S2}, z_{S2} | \xi_S = \xi_0, \eta_{S1}, l_{S1}) = \frac{1}{S_{\text{side2}}} \left\{ \iint_{S_{\text{side2}}} \frac{1}{S_{\text{side1}}} \right.$$
$$\left. \iint_{S_{\text{side1}}} \{G(\xi_r = \xi_0, \eta_{S2}, z_{S2} | \xi_S = \xi_0, \eta_{S1}, l_{S1}) h_\eta \mathrm{d}\eta_{S1} \mathrm{d}l_{S1}\} h_\eta \mathrm{d}\eta_{S2} \mathrm{d}l_{S2} \right\}. \tag{3.34}$$

In Eq. (3.34), if the side ports 2 and 1 are different, one obtains the *cross-impedance* parameter, and when response port 2 is same as the excitation port 1, one obtains the *self-impedance* parameter. In this work, we require only the self-impedance parameter expression.

The average acoustic pressure response expressions for a circular cylindrical chamber due to the end/side port excitation port and end/side port response port may be similarly derived; the reader is referred to Ref. [35] for details.

3.3 Integration of Green's Function Over the Piston Area Due to End or Side Port

3.3.1 End Port

The integrals of the product of Mathieu and modified Mathieu functions over the piston area of the end port in Eq. (3.19) are evaluated numerically, using the Simpson's three-eighths quadrature rule [32]. To this end, the integrals are converted from the (ξ_E, η_E) domain to the (x_E, y_E) domain by using the inverse Jacobian relation between the Cartesian and Elliptic coordinates [1].

$$
\iint_{S_{\text{end}}} Ce_m(\xi_E, q_{m,n}) ce_m(\eta_E, q_{m,n}) h_\xi(\xi_E, \eta_E) h_\eta(\xi_E, \eta_E) d\xi_E d\eta_E
$$

$$
= \int_{x=x_E-r_0}^{x=x_E+r_0} \int_{y=y_E-\sqrt{r_0^2-(x-x_E)^2}}^{y=y_E+\sqrt{r_0^2-(x-x_E)^2}} Ce_m(\xi_E, q_{m,n}) ce_m(\eta_E, q_{m,n}) dy_E dx_E, \qquad (3.35)
$$

and

$$
\iint_{S_{\text{end}}} Se_m(\xi_E, \bar{q}_{m,n}) se_m(\eta_E, \bar{q}_{m,n}) h_\xi(\xi_E, \eta_E) h_\eta(\xi_E, \eta_E) d\xi_E d\eta_E
$$

$$
= \int_{x=x_E-r_0}^{x=x_E+r_0} \int_{y=y_E-\sqrt{r_0^2-(x-x_E)^2}}^{y=y_E+\sqrt{r_0^2-(x-x_E)^2}} Se_m(\xi_E, \bar{q}_{m,n}) se_m(\eta_E, \bar{q}_{m,n}) dy_E dx_E. \qquad (3.36)
$$

where $\xi_E = \xi_E(x_E, y_E)$, $\eta_E = \eta_E(x_E, y_E)$ and r_0 denotes the radius of the end port.

For a circular cylindrical chamber, the reader is referred to Ref. [35] for integrals pertaining to the end port which consists of integrating the product of the Bessel functions $J_m(\cdot)$ and circular functions over the port area.

3.3.2 Side Port

The integration of the Green's function over the *elliptical* piston area is also evaluated numerically. The integral for the even and odd modes, respectively, for angular location of port center on the major-axis, i.e., $\eta_S = 0, \pi$ are shown as [1]

$$\eta_S = 0, \quad \int_{z=l_S-r_0}^{z=l_S+r_0} \left\{ \int_{\eta=-\theta_0}^{\eta=\theta_0} h_\eta ce_m(\eta, q_{m,n}) d\eta \right\} \cos\left(\frac{P\pi z}{L}\right) dz, \quad \int_{z=l_S-r_0}^{z=l_S+r_0} \left\{ \int_{\eta=-\theta_0}^{\eta=\theta_0} h_\eta se_m(\eta, q_{m,n}) d\eta \right\} \cos\left(\frac{P\pi z}{L}\right) dz,$$

$$\eta_S = \pi, \quad \int_{z=l_S-r_0}^{z=l_S+r_0} \left\{ \int_{\eta=\pi-\theta_0}^{\eta=\pi+\theta_0} h_\eta ce_m(\eta, q_{m,n}) d\eta \right\} \cos\left(\frac{P\pi z}{L}\right) dz, \quad \int_{z=l_S-r_0}^{z=l_S+r_0} \left\{ \int_{\eta=\pi-\theta_0}^{\eta=\pi+\theta_0} h_\eta se_m(\eta, q_{m,n}) d\eta \right\} \cos\left(\frac{P\pi z}{L}\right) dz,$$

$$(3.37, 3.38)$$

where

$$\sin\theta_0 = \sqrt{\left(r_0^2 - (l_S - z)^2\right)}/(D_2/2), \quad l_S - r_0 \le z \le l_S + r_0, \tag{3.39}$$

and r_0 denotes the radius of the side port. The integrals when the center of side port is located on the minor-axis, i.e., $\eta_S = \pi/2, 3\pi/2$ may be similarly evaluated; however, the limits of integration along the angular coordinates are different, as detailed below:

$$\eta_S = \pi/2: \quad \eta = \theta_S \ldots \pi - \theta_S,$$
$$\eta_S = 3\pi/2: \quad \eta = \pi + \theta_S \ldots 2\pi - \theta_S \tag{3.40}$$

where

$$\cos\theta_0 = \sqrt{\left(r_0^2 - (l_S - z)^2\right)}/(D_1/2), \quad l_S - r_0 \le z \le l_S + r_0. \tag{3.41}$$

For a circular cylindrical chamber, the reader is again referred to Ref. [35] for integrals pertaining to the side port which consists of integrating the product of circular functions over the elliptical piston area due to the side port.

It is noted that since the integrals over the end ports and side ports as well as the volume integrals $N_{m,n,P}$ and $\overline{N_{m,n,P}}$ are evaluated numerically (using a quadrature rule), this method for characterizing an elliptical (circular) chamber muffler is termed as 3-D semi-analytical approach based on the uniform piston-driven model.

3.4 Influence of Location of End and Side Ports on the Propagation of Even and Odd Modes

The frequency of excitation determines whether a transverse mode is cut-on (propagating) or is evanescent (decaying). However, even above its cut-on frequency, the propagation of a higher-order transverse mode can be suppressed by locating the port center on its pressure node [2, 36]. In other words, the angular location of the port eventually determines whether a given mode propagates or not above its cut-on frequency.

In this section, we examine the influence of the location of an end port or a side port on the possibility of propagation of the even-even, even-odd, odd-even and odd-odd modes of an elliptical chamber as well as the radial and circumferential modes of a circular cylindrical chamber [1]. Although the uniform piston-driven model, due to its higher accuracy, is eventually used to evaluate the impedance [**Z**] matrix parameters, the influence of the port location on suppression/propagation of transverse modes is most easily examined analytically through the Green's function or the point-source response, at least for automotive mufflers with port diameter considered in this monograph.

The following different cases of location of an end port are first examined through Eqs. (2.18–2.21), i.e., angular Mathieu functions and Eqs. (2.32–2.35), i.e., the hyperbolic expansion of the modified Mathieu functions.

(a) Port center is located on the major-axis at a distance greater than semi-interfocal distance, i.e., $x_E > h$ or $x_E < -h$ with $y_E = 0$,

(b) Port center is located on the focus, i.e., $x_E = h$ or $x_E = -h$ and $y_E = 0$,

(c) Port center is located on the major-axis at a distance less than the semi-inter focal distance, i.e., $|x_E| < h$ and $y_E = 0$,

(d) Port center is located at the center of the elliptical cross-section, i.e., $x_E = 0$, $y_E = 0$, and

(e) Port is located on the minor-axis, i.e., $y_E > 0$ or $y_E < 0$ and $x_E = 0$.

For the case (a), we have $\xi_E > 0$, $\eta_E = 0$ or π implying that only the even-even and even-odd modes can propagate while the odd-even and odd-odd modes do not propagate as their corresponding angular Mathieu functions evaluate to zero. For the case (b), we have $\xi_E = 0$, $\eta_E = 0$ or π, and thus, one can make the same comment on the propagation of modes as the case (a). For the case (c), $\xi_E = 0$, $0 \leq \eta_E \leq \pi$ which signifies that odd-even and odd-odd Mathieu functions evaluate to zero, implying that only the even-even and even-odd modes propagate. The case (d) is a concentric end port case: we notice that $\xi_E = 0$, $\eta_E = \pi/2$, signifying that the even-odd and odd-even Mathieu functions evaluate to zero and simultaneously, the odd-odd and odd-even modified Mathieu functions evaluate to zero. This implies that only the even-even mode propagates when the end port is located symmetrically about both the major-axis and minor-axis. Finally, for case (e), $\xi_E > 0$, $\eta_E = \pi/2$ so that the even-odd and odd-even Mathieu functions evaluate to zero so that only the even-even and odd-odd modes can propagate.

In a general case, when the end port location does not fit into either of the afore-mentioned cases, all the mode types may propagate without attenuation provided that the excitation frequency is greater than the cut-on frequency.

For the circular cylindrical chamber, if the end port is concentric, i.e., the offset distance $\delta_E = 0$, only the radial modes will propagate, whereas for an end offset port, all mode types: the radial, circumferential and cross-modes will propagate, see Eq. (3.13).

Additionally, a particular transverse mode may also be suppressed by offsetting the port center on its pressure node. For an elliptical cross-section, the $(2, 1)e$ mode can be suppressed by locating the end port center on the intersection of nodal hyper-bola and the major-axis (case (c)), refer to Fig. 2.6b. The suppression of this mode is important as it has the lowest cut-on frequency after the $(1, 1)e$ mode and elimi-nating it helps in designing a high-eccentricity elliptical muffler having an end offset port towards achieving a better attenuation performance, especially the short cham-bers, see chapter 4. Similarly, for a circular cross-section, the $(0, 1)$ radial mode is suppressed by offsetting the end port on its pressure node given by $\delta_E = 0.6276R_0$, see Ref. [25] and can help in design of a circular cylindrical muffler.

We now examine different cases of location of a side port of an elliptical chamber. First, when the port is located on the major-axis, i.e., $\eta_S = 0$ or $\eta_S = \pi$, it is observed from Eqs. (2.20) and (2.21) that $\sin(m\eta_S) = 0, \forall m = 0, 1, 2, \ldots$, therefore, the odd-even and odd-odd transverse modes are suppressed although the frequency of excitation may be greater than their cut-on frequency. On the other hand, it is observed from Eqs. (2.18) and (2.19) that the even-even and even-odd modes will always be excited above their cut-on frequency. Next, when the side port is located on the minor-axis, $\eta_S = \pi/2$ or $\eta_S = 3\pi/2$, one obtains the following conditions: (1) $\cos(m\eta_S) = 0, \forall m = 1, 3, 5, \ldots$, and (2) $\sin(m\eta_S) = 0, \forall m = 2, 4, 6, \ldots$, thereby indicating that the even-odd and odd-even modes are suppressed. On the other hand, it can be shown that the even-even and odd-even modes propagate and contribute to the total acoustic field inside the chamber when the side port is located on the minor-axis.

For the circular chamber muffler, if the relative angular location θ_0 between a side port and a side/end port is equal to $\pi/2$, the circumferential modes (and their cross-modes) corresponding to order $m = 1, 3, 5$ are suppressed.

This understanding of the influence of port location on the suppres-sion/propagation of certain mode types will be used in analyzing and designing both long and short chambers.

3.5 Evaluation of the Impedance [Z] Matrix Parameters

By virtue of the acoustic pressure response function for different combinations of piston excitation and response at the end and side port locations derived in Sect. 3.2 and the integrals presented in Sect. 3.3, one is now in a position to characterize different configurations of a single-inlet and single-outlet (SISO) mufflers through an impedance [**Z**] matrix representation [37]. On a related note, it is mentioned here that in principle, the present formulation also enables one to characterize a single-inlet and double-outlet (SIDO) elliptical cylindrical muffler [5].

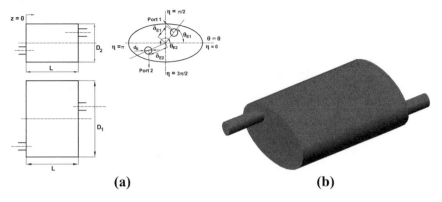

Fig. 3.3 Straight-flow elliptical expansion chamber muffler having an end-inlet port 1 and an end-outlet port 2 located on opposite end faces: **a** three orthogonal projections [35], **b** 3-D view (CAD model)

3.5.1 End-Inlet and End-Outlet Muffler

3.5.1.1 Straight-Flow Configuration

A straight-flow elliptical chamber muffler configuration is shown in Fig. 3.3 having end ports 1 and 2 located at arbitrary angular/offset location on the opposite end faces. This configuration is also popularly known as a simple expansion chamber muffler and has been the subject matter of several investigations [14, 15, 17]. This muffler configuration is characterized by the following [$\mathbf{Z}$] matrix.

$$[\mathbf{Z}] = \begin{bmatrix} p(\text{End}_1|\text{End}_1) & p(\text{End}_1|\text{End}_2) \\ p(\text{End}_2|\text{End}_1) & p(\text{End}_2|\text{End}_2) \end{bmatrix}, \tag{3.46}$$

where

$$p(\text{End}_1|\text{End}_1) = p(\xi_{E1}, \eta_{E1}, 0|\xi_{E1}, \eta_{E1}, 0),$$
$$p(\text{End}_2|\text{End}_2) = p(\xi_{E2}, \eta_{E2}, L|\xi_{E2}, \eta_{E2}, L),$$
$$p(\text{End}_2|\text{End}_1) = p(\xi_{E2}, \eta_{E2}, L|\xi_{E1}, \eta_{E1}, 0) = p(\text{End}_1|\text{End}_2). \tag{3.47a–c}$$

3.5.1.2 Flow-Reversal Configuration

A flow-reversal elliptical chamber muffler configuration has both end ports located on the same end face, in general, at arbitrary angular/offset locations, see Fig. 3.4.

As noted previously, the flow-reversing chamber is an important component of the exhaust mufflers in heavy-duty transportation vehicles and is often used as an

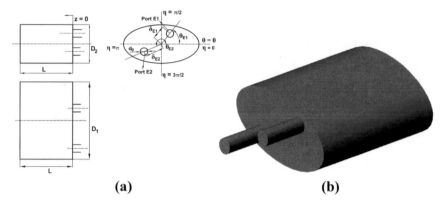

Fig. 3.4 Flow-reversal elliptical expansion chamber muffler having an end-inlet port 1 and an end-outlet port 2 located on the same end face: **a** three orthogonal projections, **b** 3-D view (CAD model). The schematic diagrams are adopted from Ref. [7]

end-chamber in multi-pass perforated silencer system [38]. For this configuration, the higher-order modes generated at a port (area discontinuities) of these silencers do not decay fully before reaching the other port due to their close proximity [7, 18, 25]. This muffler configuration is also characterized by Eq. (3.46) wherein

$$p\left(\text{End}_j \big| \text{End}_i\right) = p\left(\xi_{Ej}, \eta_{Ej}, 0 \big| \xi_{Ei}, \eta_{Ei}, 0\right), \quad \forall j, i = 1, 2. \tag{3.48}$$

The diameter of both the end ports in muffler configurations shown in Figs. 3.3 and 3.4 is taken equal and is denoted by d_0. Their location on the elliptical cross-section is specified in terms of the offset distances δ_{E1}, δ_{E2}, respectively, measured with reference to the center, and the angular location θ_{E1}, θ_{E2}, respectively, measured with respect to $\eta = 0$ axis, refer to part (a) of Figs. 3.3 and 3.4. Note that δ and θ are, in general, related to the radial elliptical ξ and angular elliptical η coordinates by the following relationships.

$$\delta = h\sqrt{\cosh^2 \xi - \sin^2 \eta}, \tag{3.49}$$

and

$$\theta = \cos^{-1}\left\{ \frac{\cosh \xi \cos \eta}{\sqrt{\cosh^2 \xi - \sin^2 \eta}} \right\} = \sin^{-1}\left\{ \frac{\sinh \xi \sin \eta}{\sqrt{\cosh^2 \xi - \sin^2 \eta}} \right\}. \tag{3.50}$$

3.5.2 End-Inlet and Side-Outlet Configuration

An elliptical muffler configuration having an end (side) inlet port and side (end) outlet port is shown in Fig. 3.5. The end port 1 is located at an arbitrary angular/offset location, while the side port is located at a generic angle $\eta = \eta_S$ and axial distance $z = l_S$ (equal diameters are considered for both ports.). However, in Chap. 4, we will analyze only those configurations where the side port location is located on the major-axis, i.e., at $\eta = 0, \pi$.

This end-inlet side-outlet configuration shown in Fig. 3.5 is characterized using the following $[\mathbf{Z}]$ matrix.

$$[\mathbf{Z}] = \begin{bmatrix} p(\text{End}|\text{End}) & p(\text{End}|\text{Side}) \\ p(\text{Side}|\text{End}) & p(\text{Side}|\text{Side}) \end{bmatrix}, \tag{3.51}$$

where

$$p(\text{Side}|\text{End}) = p(\xi_S, \eta_S, l_S | \xi_E, \eta_E, l_E) = p(\text{End}|\text{Side}), \tag{3.52}$$

and

$$p(\text{Side}|\text{Side}) = p(\xi_S, \eta_S, l_S | \xi_S, \eta_S, l_S). \tag{3.53}$$

The analysis of the configurations shown in Fig. 3.5 is important because sometimes, based on purely logistical reasons, the inlet and/or outlet ports must be located on the sides with their axes normal to the chamber axial direction. In fact, the analysis of these end-inlet and side-outlet configurations has been reported in previous studies

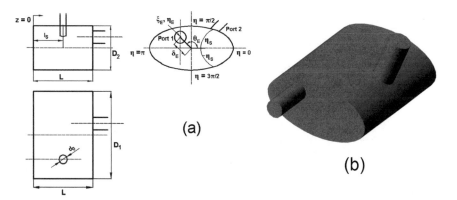

Fig. 3.5 Elliptical cylindrical expansion chamber muffler having an end-inlet (outlet) port 1 and side-outlet (inlet) port 2: **a** three orthogonal projections, **b** 3-D view (CAD model). The schematic diagrams are adopted from Ref. [1]

based on either, the 1-D axial plane wave model [39, 40] or a three-dimensional analytical/numerical approach for circular [19, 20] and elliptical [1, 5] cross-sectional geometries. Furthermore, since the convective effect of mean flow is not considered and rigid-wall empty chambers (without dissipative material) are analyzed, we are essentially considering a reciprocal and conservative system [11].

Note here that the [Z] matrix parameters are in essence, the same as the *frequency response function* computed by Denia et al. [6] for an elliptical chamber muffler with end ports based on the 3-D analytical point-source modeling. While they also obtained the [Z] matrix, it was converted to the transfer [T] matrix using the well-known relation between the [Z] and [T] matrix parameters [11]. These [T] matrix parameters, also known as the four-pole parameters [2], can be obtained through the following relations:

$$T_{11} = \left.\frac{p_1}{p_2}\right|_{v_2=0} = \frac{f_1(\xi_1, \eta_1, z_1)}{f_1(\xi_2, \eta_2, z_2)}, \quad T_{21} = \left.\frac{v_1}{p_2}\right|_{v_2=0} = \frac{1}{f_1(\xi_2, \eta_2, z_2)}, \quad (3.54, 3.55)$$

$$T_{22} = v_1/v_2|_{p_2=0} = -f_2(\xi_2, \eta_2, z_2)/f_1(\xi_2, \eta_2, z_2), \quad (3.56)$$

and

$$T_{12} = \left.\frac{p_1}{v_2}\right|_{p_2=0} = f_2(\xi_1, \eta_1, z_1) - \left(\frac{f_2(\xi_2, \eta_2, z_2)}{f_1(\xi_2, \eta_2, z_2)}\right) f_1(\xi_1, \eta_1, z_1). \quad (3.57)$$

Here, $f_1(\cdot)$ and $f_2(\cdot)$ are the pressure response functions, and they related the acoustic pressure and mass velocities at the end-inlet ports 1 and 2 as

$$p_1 = \underbrace{f_1(\xi_1, \eta_1, z_1)}_{Z_{11}} v_1 + \underbrace{f_2(\xi_1, \eta_1, z_1)}_{Z_{12}} v_2, \quad (3.58)$$

and

$$p_2 = \underbrace{f_1(\xi_2, \eta_2, z_2)}_{Z_{21}} v_1 + \underbrace{f_2(\xi_2, \eta_2, z_2)}_{Z_{22}} v_2, \quad (3.59)$$

respectively. It is noted that following Ref. [37], the direction of mass velocity is considered positive 'looking into the system' at both ports in the present formulation. The transmission loss (TL) performance of a SISO muffler can now be evaluated in terms of the [T] matrix parameters. However, in this monograph, we will evaluate TL performance directly in terms of the [Z] matrix parameters avoiding the need to first obtain the [T] matrix.

3.6 Computation of Transmission Loss Performance in Terms of [Z] Matrix Parameters

An expression for the transmission loss (TL) performance of a SISO system is obtained here in terms of the [Z] matrix parameters. To this end, the relation between the scattering [S] matrix and the [Z] matrix is first presented [37].

$$\left\{ \begin{matrix} B_1 \\ B_2 \end{matrix} \right\} = [S] \left\{ \begin{matrix} A_1 \\ A_2 \end{matrix} \right\}, \quad [S]_{2\times2} = \begin{bmatrix} S_{11} & S_{12} \\ S_{21} & S_{22} \end{bmatrix}, \quad [S] = \mathbf{I} - 2 \left[[Z] \begin{bmatrix} 1/Y_1 & 0 \\ 0 & 1/Y_2 \end{bmatrix} + \mathbf{I} \right]^{-1},$$

$$(3.60\text{a--c})$$

where $\{B_1, B_2\}$ and $\{A_1, A_2\}$ are the reflected progressive-wave and the incident progressive-wave amplitudes, respectively, at the end ports, $\mathbf{I}$ is the identity matrix while Y_1 and Y_2 are the characteristic impedances at the ports 1 and 2, respectively. In obtaining the [S] matrix, it is assumed that only planar waves or axial wave propagation exist in the end ports from the port-chamber interface, although a 3-D field is considered inside the muffler chamber. Therefore, as discussed in Sect. 3.2, this assumption essentially considers the ports as rigid oscillating pistons having uniform velocity equal to the normal acoustic particle velocity in the muffler chamber over the port cross-sectional area.

A uniform piston excitation is applied at the port E1, while anechoic termination is imposed at the port E2, thereby implying $A_2 = 0$. The TL is, therefore, given by

$$\text{TL} = 10 \log_{10} \left(\frac{Y_2}{Y_1} \frac{1}{|S_{21}|^2} \right) = 10 \log_{10} \left(\frac{1}{4Y_1Y_2} \left| \frac{(Z_{11} + Y_1)(Z_{22} + Y_2) - Z_{21}Z_{12}}{Z_{21}} \right|^2 \right).$$

$$(3.61)$$

The TL expression remains unaltered for piston excitation at port E2 and anechoic termination condition imposed at port E1 as a reciprocal SISO system is considered, i.e., $Z_{12} = Z_{21}$.

3.6.1 Analysis/Prediction of the Observed Peaks and Troughs in the TL Spectrum

The advantage of expressing the TL expression for a reciprocal SISO muffler system explicitly in terms of the [Z] matrix parameters given by Eq. (3.61) is that it enables one to readily analyze/predict the characteristic features of the TL spectrum such as the frequency of occurrence of the attenuation peaks and the troughs in a rather straightforward manner. A peak in the TL graph of a general SISO muffler occurs at a frequency f when either of the following conditions is satisfied by the [Z] matrix parameters [7, 22]:

(a) $Z_{11} \to \infty$, while Z_{22}, Z_{21} (and Z_{12}) are finite,
(b) $Z_{22} \to \infty$, while Z_{11}, Z_{21} (and Z_{12}) are finite, and
(c) $Z_{11} \to \infty$ and $Z_{22} \to \infty$, while Z_{21} (and Z_{12}) are finite,
(d) $Z_{21} = 0$ (or equivalently, $Z_{12} = 0$), regardless of the order of magnitudes of the self-impedance parameters Z_{11} and Z_{22}.

It is noted that since a reciprocal muffler system is considered here, the following conditions will hold good at a given frequency f: (1) if Z_{21} is finite, it implies that Z_{12} is also finite and (2) $Z_{21} = 0 \Rightarrow Z_{12} = 0$. When conditions (a–c) are satisfied, the numerator of the RHS of Eq. (3.61) tends to infinity, while when condition (d) is satisfied, its denominator tends to zero, implying a large attenuation or peak at f in the TL spectrum of a reciprocal muffler system.

It is worth mentioning here that if mean flow is nonzero, the SISO muffler does not satisfy acoustic reciprocity, i.e., $Z_{21} \neq Z_{12}$. Nevertheless, the TL spectrum will still exhibit an attenuation peak at a frequency f when (1) conditions (a–d) are satisfied. Of course, for such muffler system, Z_{21} and Z_{12} must both be finite (conditions a–c) or $Z_{21} = 0$ (condition d) as $Z_{12} = 0$ alone does not guarantee a peak.

A trough in the TL graph of a SISO muffler occurs at a given frequency f when all the [**Z**] matrix parameters that are purely imaginary for a reciprocal and conservative system (as is the case here) tend to infinity, i.e., $Z_{ij}(f) \to \infty$ for $i, j = 1, 2$ at the rate $O(\varepsilon^{-2})$, $\varepsilon \to 0$. This implies $Z_{11}Z_{22} - Z_{12}Z_{21} \approx 0$ in numerator of Eq. (3.63). Thus, under this condition, TL expression simplifies to

$$\text{TL} \approx 10 \log_{10} \left(\frac{1}{4Y_1 Y_2} \left| \frac{Z_{11}Y_2 + Z_{22}Y_1}{Z_{21}} \right|^2 \right) \approx 10 \log_{10} \left(\frac{(Y_1 + Y_2)^2}{4Y_1 Y_2} \right), \quad (3.62)$$

and for ports with equal diameters, i.e., $Y_1 = Y_2$, the TL is exactly equal to zero, thereby indicating the occurrence of a trough.

References

1. A. Mimani, M.L. Munjal, 3-D acoustic analysis of elliptical chamber mufflers having an end inlet and a side outlet: an impedance matrix approach. Wave Motion **49**, 271–295 (2012)
2. M.L. Munjal, *Acoustics of Ducts and Mufflers*, 2nd edn. (Wiley, Chichester, UK, 2014)
3. E. Ramya, M.L. Munjal, Improved tuning of the extended concentric tube resonator for wideband transmissions loss. Noise Control Eng. J. **62**, 252–263 (2014)
4. N.W. McLachlan, *Theory and Application of Mathieu Functions* (Oxford University Press, London, 1947)
5. A. Mimani, M.L. Munjal, Acoustical behavior of single inlet and multiple outlet elliptical cylindrical chamber muffler. Noise Control Eng. J. **60**, 605–626 (2012)
6. F.D. Denia, J. Albelda, F.J. Fuenmayor, A.J. Torregrosa, Acoustic behaviour of elliptical chamber mufflers. J. Sound Vibr. **241**, 401–421 (2001)
7. A. Mimani, M.L. Munjal, Acoustic end-correction in a flow-reversal end chamber muffler: a semi-analytical approach. J. Comput. Acoust. **24**, 1650004 (2016)

8. Y.-H. Kim, S.-W. Kang, Green's solution of the acoustic wave equation for a circular expansion chamber with arbitrary locations of inlet, outlet port, and termination impedance. J. Acoust. Soc. Am. **94**, 473–490 (1993)
9. W. Shao, C.K. Mechefske, Acoustic analysis of a finite cylindrical duct based on Green's functions. J. Sound Vibr. **287**, 979–988 (2005)
10. L.E. Kinsler, A.R. Frey, A.B. Coppens, J.V. Sanders, Fundamentals of Acoustics. (Wiley, New York, 2000), pp. 172–176
11. V. Easwaran, V.H. Gupta, M.L. Munjal, Relationship between the impedance matrix and the transfer matrix with specific reference to symmetrical, reciprocal and conservative systems. J. Sound Vib. **161**, 515–525 (1993)
12. J. Kim, W. Soedel, General formulation of four pole parameters for three-dimensional cavities utilizing modal expansion, with special attention to the annular cylinder,. J. Sound Vib. **129**, 237–254 (1989)
13. P.C.-C. Lai, W. Soedel, Two-dimensional analysis of thin, shell or plate like muffler elements. J. Sound Vib. **194**, 137–171 (1996)
14. A. Selamet, P.M. Radavich, The effect of length on acoustic attenuation performance of concentric expansion chambers: an analytical, computational and experimental investigation. J. Sound Vib. **201**, 407–426 (1997)
15. A. Selamet, Z.L. Ji, Acoustic attenuation performance of circular expansion chambers with offset inlet/outlet: I. Analytical approach. J. Sound Vib. **213**, 601–617 (1998)
16. W. Zhou, J. Kim, Formulation of four poles of three-dimensional acoustic systems from pressure response functions with special attention to source modelling. J. Sound Vib. **219**, 89–103 (1999)
17. J.-G. Ih, B.-H. Lee, Analysis of higher order mode effects in circular expansion chamber with mean flow. J. Acoust. Soc. Am. **77**, 1377–1388 (1985)
18. J.-G. Ih, B.-H. Lee, Theoretical prediction of the transmission loss for the circular reversing chamber mufflers. J. Sound Vib. **112**, 261–272 (1987)
19. S.I. Yi, B.-H. Lee, Three-dimensional acoustic analysis of a circular expansion chamber with side inlet and side outlet. J. Acoust. Soc. Am. **79**, 1299–1306 (1986)
20. S.I. Yi, B.-H. Lee, Three-dimensional acoustic analysis of a circular expansion chamber with side inlet and end outlet. J. Acoust. Soc. Am. **81**, 1279–1287 (1987)
21. H. Keskar, B. Venkatesham, Transmission loss characteristics of an annular cavity with arbitrary port locations using Green's function method. J. Acoust. Soc. Am. **142**, 1350–1361 (2017)
22. A. Mimani, M.L. Munjal, Design of reactive rectangular expansion chambers for broadband acoustic attenuation performance based on optimal port location. Acoust. Aust. **44**, 299–323 (2016)
23. J.-G. Ih, The reactive attenuation of rectangular plenum chambers. J. Sound Vib. **157**, 93–122 (1992)
24. A. Selamet, Z.L. Ji, Acoustic attenuation performance of circular expansion chambers with extended inlet/outlet. J. Sound Vibr. **223**, 197–212 (1999)
25. A. Selamet, Z.L. Ji, Acoustic attenuation performance of circular flow-reversing chambers. J. Acoust. Soc. Am. **104**, 2867–2877 (1998)
26. F.D. Denia, L. Baeza, J. Albelda, F.J. Fuenmayor, Acoustic behaviour of elliptical mufflers with single-inlet and double-outlet, in *Proceedings of the 10th International Congress of Sound and Vibration*, 7–10 July 2003, Stockholm, Sweden
27. R. Kirby, Transmission loss predictions for dissipative silencers of arbitrary cross section in the presence of mean flow. J. Acoust. Soc. Am. **114**, 200–209 (2003)
28. J. Albelda, F.D. Denia, M.I. Torres, F.J. Fuenmayor, A transversal substructuring mode matching method applied to the acoustic analysis of dissipative mufflers. J. Sound Vibr. **303**, 614–631 (2007)
29. F.D. Denia, E.M. Sánchez-Orgaz, J. Martínez-Casas, R. Kirby, Finite element based acoustic analysis of dissipative silencers with high temperature and thermal-induced heterogeneity. Finite Elem. Anal. Des. **101**, 46–57 (2015)
30. Z. Fang, Z.L. Ji, Numerical mode matching approach for acoustic attenuation predictions of double-chamber perforated tube dissipative silencers with mean flow. J. Comput. Acoust. **22**, 1450004 (2014)

31. T. Myint-U, L. Debnath, *Linear Partial Differential Equations for Scientists and Engineers.* (Birkhäuser, Boston, 2007), pp. 430–432
32. Erwin Kreyszig, *Advanced Engineering Mathematics*, 10th edn. (Wiley, New Jersey, USA, 2011)
33. A. Mimani, M.L. Munjal, Transverse plane wave analysis of short elliptical chamber mufflers: an analytical approach. J. Sound Vib. **330**, 1472–1489 (2011)
34. A. Mimani, M.L. Munjal, On the role of higher-order evanescent modes in end-offset inlet and end-centered outlet elliptical flow-reversal chamber mufflers. Int. J. Acoust. Vib. **17**, 139–154 (2012)
35. A. Mimani, 1D and 3D analysis of multi-port muffler configurations with emphasis on elliptical cylindrical chamber. PhD Thesis, Indian Institute of Science, Bangalore (2012)
36. L.J. Eriksson, Effect of inlet/outlet locations on higher order modes in silencers. J. Acoust. Soc. Am. **72**, 1208–1211 (1982)
37. A. Mimani, M.L. Munjal, Acoustical analysis of a general network of multi-port elements—an impedance matrix approach. Int. J. Acoust. Vib. **17**, 23–46 (2012)
38. N.S. Dickey, A. Selamet, K.V. Tallio, Multi-pass perforated tube silencers: a computational approach. J. Sound Vibr. **211**, 435–447 (1999)
39. M.L. Munjal, Plane wave analysis of side inlet/outlet chamber mufflers with mean flow. Appl. Acoust. **52**, 165–175 (1997)
40. K.M. Kumar, M.L. Munjal, On development of rational design guidelines for large side-inlet side-outlet perforated element mufflers. Noise Control Eng. J. **66**, 308-323 (2018)

Chapter 4
Double-Tuned Short End-Chamber Mufflers

4.1 Multi-pass Perforated Duct Muffler System

Multi-pass perforated (MPP) tube mufflers are widely used in modern-day automotive exhaust systems to reduce engine noise. Such muffler configurations are characterized by a good attenuation performance primarily due to the interaction of the acoustic waves propagating in the perforated tubes with those in the annular cavity and are popular because they can deliver large attenuation within the constraints of a limited space beneath the automobile. However, this comes at the expense of relatively greater back-pressure than straight-through flow mufflers. As many of the commercial mufflers implement three passes, i.e., three perforated ducts housed within an expansion chamber, such configurations have been a subject matter of several investigations [1–10]. The first published work on the analysis of the three-pass muffler was by Ross [1] who used finite-element analysis (FEA). However, the results were demonstrated only for the configuration in which the middle pipe was just pass through, i.e., not perforated. His results showed good agreement with measurements at low frequencies. Dickey et al. [2] developed the 1-D time-domain computational analysis of a MPP muffler and illustrated the results for a three-pass perforated (TPP) system. The corresponding frequency-domain 1-D analysis was presented by Munjal [3] based on the axial plane wave model via a decoupling method whereby the overall [**T**] matrix and TL performance of the TPP muffler configuration was obtained. A more general formulation, however, allowed analyzing tube extensions in the end-chambers [4]. Selamet et al. [5] presented a quasi-1-D approach to analyze TPP mufflers without/with duct extensions into the end-chamber cavity. Their approach was validated by comparisons with the experimental results on a few fabricated and commercially available configurations. They investigated the effect of porosity, expansion chamber diameter, porosity, length of the end-chambers along with the length of ducts protruding into the end-chambers. Ji and Selamet [6] further extended this work by proposing a substructure boundary element method (BEM) to predict the TL performance and conducted parametric studies on the TPP muffler configuration. Ji and Fang [7] employed the plane wave analysis to compute the

A. Mimani, *Acoustic Analysis and Design of Short Elliptical End-Chamber Mufflers*, https://doi.org/10.1007/978-981-10-4828-9_4

attenuation performance of TPP mufflers and is the standard practise, their results were validated by comparison with FEA and BEM predictions. It was found that the presence of an end-resonator can enhance the attenuation performance at low frequencies. Fan and Ji [8] also used the 1-D model to study the effect of perforated bulkheads on TL performance wherein it was found that increase in porosity of the bulkhead can lead to a shift in the resonance towards the higher frequencies. (Bulkheads are the common end face between the middle chamber and the end-chambers.) Additionally, they found that filling a sound-absorbing material in the annular region of the expansion chamber can improve the performance from mid-to-high frequencies, while it is somewhat lowered near the resonance frequency, thereby producing a flatter but a broadband attenuation performance for such hybrid TPP mufflers. Huang et al. [9] also investigated the effect of sound-absorbing material and porosity of the tubes and bulkheads on the attenuation performance but used the 3-D FEA which allowed them to visualize the mode shape of the first higher-order transverse mode pertaining to different elliptical cross-sections of the TPP muffler. Verma and Munjal [10] studied the effects of absorptive material, porosity of the baffle (end) plates and perforated tubes and the perforation length on the attenuation performance of a three-chamber U-bend hybrid muffler by means of 1-D transfer matrices.

4.1.1 Short End-Chambers: Integral Component of a Multi-pass Perforated Muffler

A defining feature of the TPP muffler (or for that matter, a MPP muffler) is the end-chamber which is located either at both front and rear, or only at the rear part. Logistically, the main purpose of an end-chamber is to facilitate the reversal or re-orientation of the mean flow direction, thereby forcing the flow to pass through multiple perforated tubes while the acoustic waves can interact with the annular cavity of the middle chamber resulting in high levels of attenuation. The interior view of a typical TPP muffler having an elliptical cross-section was previously shown in Fig. 1.1b where the important constituting elements including short end-chambers on both sides of the middle chamber were shown. Space constraints along the axial direction allow the use of end-chambers with only a short length where the length-to-diameter ratio is generally close to 0.2 or so.

While the aforementioned papers, especially the ones using 3-D numerical approaches do a reasonably good job insofar as the overall acoustic analysis and design measures are considered, there is almost a complete lack of guidelines in the literature for optimal acoustic design of short flow-reversal end-chambers, although some work is reported for such configuration having a longer length [11–13]. By 'optimal' acoustic design, we mean determining specific locations or particular arrangements of inlet and outlet ports of the short end-chamber which can deliver a broadband attenuation which covers a maximum possible frequency range. Alternatively, one can also design to achieve a relatively high attenuation in a desired frequency range, possibly at the expense of a not so good performance or breakdown outside this range. In this chapter, we focus on the former design objective, i.e., achieving a flatter but a broadband attenuation pattern for a given location of inlet and outlet. This problem is important because ignoring or overlooking the design of end-chamber can potentially have an adverse effect on the overall TL performance of the TPP muffler configurations. We must however, immediately mention that although limited papers which present some perspectives on the short chamber design do exist, they employ a multi-dimensional (3-D) model and suggest specific configurations which include an end-centered and end-inlet flow-reversal short circular chamber [14] and an end-inlet and side-outlet short elliptical chambers [15, 16] that deliver a wideband performance. Note that due to the short length, the higher-order evanescent modes that are generated at the port-chamber interface do not decay sufficiently and in fact, completely dominate the chamber modes, even in the low-frequency range! While this makes it necessary to use the 3-D (or 2-D) analytical approaches [11, 14, 15, 17] or use FEA [12, 16], simpler mathematical models have indeed been proposed which consider the plane wave propagation along the major-axis [18, 19] or along the diameter [20] for short chambers of elliptical and circular cross-sections, respectively.

The primary objective of this chapter is then to present a consolidated analysis of short elliptical and circular end-chambers with different orientation of inlet and outlet ports that can deliver the best attenuation performance within the constraints of limited space. The short end-chamber mufflers analyzed in this chapter include the end-inlet and end-outlet configurations of the flow-reversal or straight-flow type as well as the end-inlet and side-outlet configuration. The possible use of these short elliptical end-chamber configurations is illustrated through the schematics shown in Fig. 4.1a–e. This chapter brings together the different models for analysis, namely the 3-D and 2-D semi-analytical approaches as well as the simpler 1-D transverse plane wave approach to achieve the desired objective, and the analytical results are validated by comparison against the 3-D FEA for specific cases.

4.2 End-Inlet and End-Outlet Configurations

Figure 4.2a shows the 3-D CAD model of the short end-chamber elliptical muffler with ports located on the same end face, i.e., a flow-reversal configuration, while Fig. 4.2b shows the 3-D model of the short end-chamber muffler with ports located on opposite end faces, i.e., a straight-flow configuration. Note that the three orthogonal projections of the straight-flow and flow-reversal configurations were previously shown in Figs. 3.3a and 3.4a, respectively. The inlet and outlet ports are centered at a generic location on the end face; their radial coordinates are given by $(\delta_{E1}, \theta_{E1})$ and $(\delta_{E2}, \theta_{E2})$, respectively, measured from the center of the elliptical section as shown in part (a) of Figs. 3.3 and 3.4. Figure 4.2c depicts the end face of an elliptical muffler showing different configurations for the angular location of end-inlet and end-outlet ports.

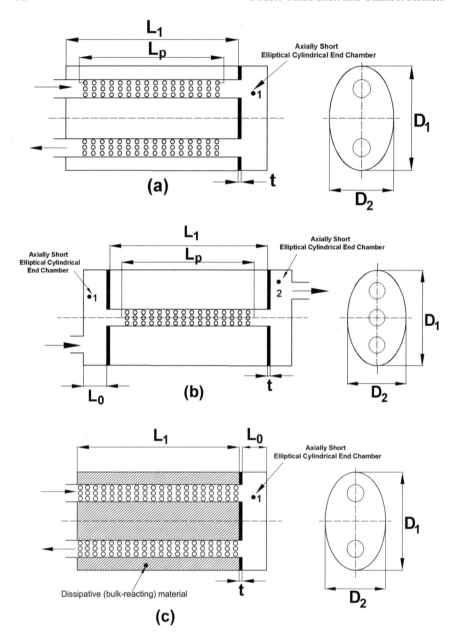

Fig. 4.1 Schematics illustrating the use of short elliptical end-chamber in two-perforated-tube-airway single flow-reversal muffler system (parts **a** and **c**) as well as in one-perforated-tube-airway straight-flow muffler system (parts **b**, **d** and **e**). In muffler systems (**a**), (**b**) and (**e**), the annular resonator cavity is empty, while in muffler systems (**c**) and (**d**), it is filled with dissipative (absorptive) material. In part (**e**), the short end-chamber on the right has an end-inlet and side-outlet ports

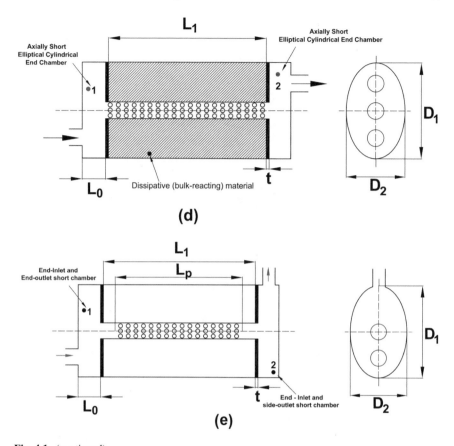

Fig. 4.1 (continued)

4.2.1   Inlet on the Major-Axis and Outlet on the Minor-Axis: Optimal Port Location

Short flow-reversal and straight-through elliptical muffler having an end-inlet port 1 centered on the intersection of the pressure nodal hyperbola $\eta_{(2,1)e}$ of the $(2, 1)e$ mode and the major-axis, and end-outlet port 2 centered on the intersection of the pressure nodal ellipse $\xi_{(0,2)e}$ of the $(0, 2)e$ mode and the minor-axis is considered. More precisely, the end port 1 is located at offset distance on the major-axis ($\theta_{E1} = 0$) given by the non-dimensional relation

$$\frac{\delta_{E1}}{D_1/2} = \frac{x_{(2,1)e}}{D_1/2} = e \cos \eta_{(2,1)e}, \tag{4.1}$$

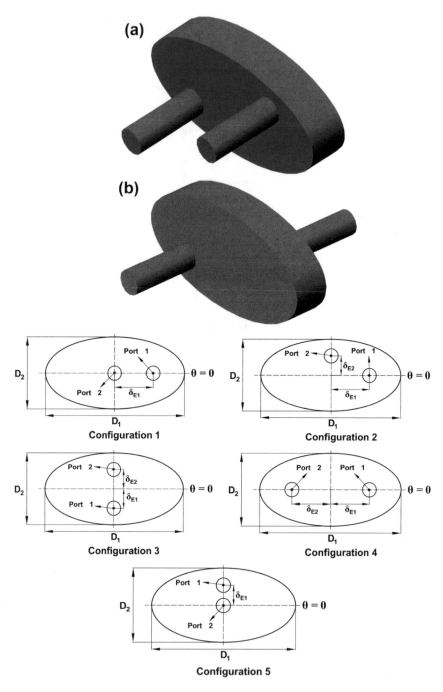

Fig. 4.2 Parts **a** and **b**: Three-dimensional view (CAD model) of short elliptical end-chamber: **a** flow-reversal and **b** straight-flow configuration, **c** end face of an elliptical cylindrical muffler showing different angular locations of the end-inlet port 1 and the end-outlet port 2 (adopted from Ref. [11]). The analysis in the ensuing section is primarily focused on the configuration 2

while the end port 2 is located at offset distance on the minor-axis ($\theta_{E2} = \pi/2$) given by the non-dimensional relation

$$\frac{\delta_{E2}}{D_1/2} = \frac{y_{(0,2)e}}{D_1/2} = e \sinh \xi_{(0,2)e}, \tag{4.2}$$

and this location of end ports is illustrated by the *configuration 2* in Fig. 4.2c. The effectiveness of this optimal port configuration towards achieving a broadband acoustic attenuation is investigated in Sect. 4.2.1. To this end, elliptical sections having major-axis $D_1 = 250$ mm and chamber length fixed at $L = 50$ mm, i.e., short chambers with $L/D_1 = 0.2$ are considered as default values throughout this chapter, while the aspect-ratio D_2/D_1 is varied. Therefore, for a circular chamber, the diameter $D_0 = D_1 = 250$ mm is considered unless otherwise stated. Furthermore, in this chapter, the port diameters equal to 40 mm are considered, and room temperature $T_0 = 20°$ is assumed implying a uniform sound speed $c_0 = 343.23$ m·s^{-1}. Furthermore, unless otherwise stated, the TL is computed up to a maximum frequency $f_0 = 3500$ Hz which is sufficient to cover the entire engine noise spectrum spanning the firing frequency and its first few integral multiples. The results, however, are shown over a non-dimensional frequency or Helmholtz number range given by $[0, 0.5k_0D_1 = 8]$. Furthermore, in the acoustic pressure response function obtained in Eq. (3.6) used for computing the $[\mathbf{Z}]$ matrix parameters for elliptical mufflers, the first six orders of the even-even modes $Ce_{2m}(\xi, q_{2m,n})ce_{2m}(\eta, q_{2m,n})$, even-odd modes $Ce_{2m-1}(\xi, q_{2m-1,n})ce_{2m-1}(\eta, q_{2m-1,n})$, odd-even modes $Se_{2m}(\xi, \overline{q}_{2m,n})se_{2m}(\eta, \overline{q}_{2m,n})$ and odd-odd modes $Se_{2m-1}(\xi, \overline{q}_{2m-1,n})se_{2m-1}(\eta, \overline{q}_{2m-1,n})$ were considered, and for each order and mode-type, the first six roots, i.e., $n = \{1, 2, \ldots, 6\}$ along with the first ten axial modes, i.e., $P = \{0, 1, \ldots, 10\}$ were considered. Similarly, for circular chamber mufflers, the first ten orders $m = \{0, 1, \ldots, 9\}$ of the Bessel function, the first ten roots, i.e., $n = \{0, 1, \ldots, 9\}$ and the first ten axial modes in Eq. (3.13) were considered for computing the $[\mathbf{Z}]$ matrix parameters. The truncation of modal summation is found to be sufficient to ensure convergence over the entire frequency range of interest and thus, is used throughout this chapter [11, 15].

Figure 4.3a and b presents the transmission loss (TL) graphs of short flow-reversal and straight-flow elliptical chambers, respectively, having an aspect-ratio $D_2/D_1 = 0.5$ or a high eccentricity given by $e = 0.8660$. Note that the vertical solid lines in Fig. 4.3 indicate the cut-on or resonance frequencies of the higher-order transverse modes of the elliptical (circular) chamber—the same convention is followed for the remaining figures in this chapter. The end-inlet port 1 is located on the major-axis with its center at offset distance $\delta_{E1} = 52.98$ mm, while the end-outlet port 2 is located on the minor-axis with its center at offset distance $\delta_{E2} = 31.99$ mm. Table 4.1 documents the values of pressure nodal hyperbola $\eta_{(2,1)e}$ and ellipse $\xi_{(0, 2)e}$ of the $(2, 1)e$ and $(0, 2)o$ modes, respectively, and the corresponding non-dimensional offset distances for $D_2/D_1 = 0.5$.

Recall here that the uniform piston-driven model [11, 15] replaces the inlet and outlet ports as rigid oscillating pistons which implies that the modal coupling between

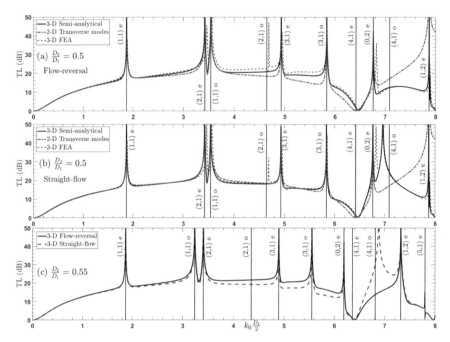

Fig. 4.3 TL performance of short **a** flow-reversal and **b** straight-flow elliptical end-chamber configurations having $D_2/D_1 = 0.5$. The 3-D semi-analytical approach is corroborated by comparing against the 3-D FEA results. **c** TL performance of short flow-reversal and straight-flow elliptical end-chamber configurations having $D_2/D_1 = 0.55$. In **a**–**c**, end-inlet and end-outlet ports are located on the major-axis ($\delta_{E1} = x_{(2,1)e}$) and minor-axis ($\delta_{E2} = y_{(0,2)e}$), respectively

the chamber and port modes is inherently not considered. Rather, the assumption of plane wave propagation immediately from the port-chamber interface implies only a one-way coupling. In practise, however, the modal coupling may be important, especially when the port diameter is large, therefore, previous papers have considered analytical or numerical mode-matching schemes which is naturally more accurate but far more tedious [12, 14]. Here, we demonstrate the accuracy of the piston-driven model by comparing the results with those obtained using the complete 3-D FEA in which the chamber and ports are considered as an integrated system, and they are meshed using the ten-noded tetrahedral elements [21]. This implies that the 3-D FEA inherently considers the modal coupling between the chamber and ports and can be used to assess the accuracy of the piston-driven model.

Figure 4.3a and b shows that, in an overall sense, an excellent agreement is observed between the uniform piston-driven model and 3-D FEA results computed using the commercial software SYSNOISE, thereby validating the former approach.[1] Both configurations exhibit a broadband attenuation performance throughout the

---

[1]In the author's previous publications [11, 15, 22], the 3-D semi-analytical uniform piston-driven model was used to compute the TL performance of (axially) long elliptical chamber automotive mufflers where the analytical framework allows the location of inlet and outlet ports either on the end

Table. 4.1 Pressure nodes of (2, 1) even and (0, 2) radial modes of the elliptical cross-section and non-dimensional frequency range over which a broadband attenuation is achieved for short elliptical flow-reversal/straight-flow chamber muffler for a range of aspect-ratio D_2/D_1

D_2/D_1	Pressure node of (2, 1) even mode $\left\{\begin{array}{l}\xi_{(2,1)e}=0, \\ y_{(2,1)e}=0\end{array}\right\}$		Pressure node of (0, 2) even mode $\{x_{(0,2)e}=0\}$		Non-dimensional frequency range of broadband attenuation		Maximum ratio of the chamber length to the major-axis: $\frac{L}{D_1} \leq \frac{\pi}{2\alpha}$
	$\eta_{(2,1)e}$	$\frac{x_{(2,1)e}}{D_1/2}$	$\xi_{(0,2)e}$	$\frac{y_{(0,2)e}}{D_1/2}$	Up to	$0 \leq k_0 \frac{D_1}{2} \leq \alpha$	
0.35	1.0893	0.4338	0.1875	0.1766	(4, 1) Even mode	$\alpha = 6.5421$	0.2401
0.40	1.0810	0.4312	0.2193	0.2026	(4, 1) Even mode	$\alpha = 6.5077$	0.2414
0.45	1.0711	0.4279	0.2557	0.2290	(4, 1) Even mode	$\alpha = 6.4669$	0.2429
0.50	1.0594	0.4238	0.2914	0.2559	(4, 1) Even mode	$\alpha = 6.4180$	0.2448
0.5231	1.0509	0.4206	0.3177	0.2735	(4, 1) Even mode	$\alpha = 6.3816$	0.2461
0.55	1.0457	0.4187	0.3332	0.2835	(4, 1) Even mode	$\alpha = 6.3593$	0.2470
0.60	1.0297	0.4121	0.3807	0.3120	(4, 1) Even mode	$\alpha = 6.2885$	0.2498
0.65	1.0110	0.4036	0.4357	0.3417	(4, 1) Even mode	$\alpha = 6.2032$	0.2532
0.70	0.9891	0.3924	0.5010	0.3729	(4, 1) Even mode	$\alpha = 6.1024$	0.2574
0.7357	**0.9713**	**0.3822**	**0.5562**	**0.3965**	**(2, 2) Even mode** $\{q_{(4,1)e} = q_{(1,2)e} = 4.1579\}$	$\alpha = 7.3649$	**0.2133**
0.75	0.9637	0.3774	0.5810	0.4063	(2, 2) Even mode	$\alpha = 7.3127$	0.2148
0.80	0.9343	0.3566	0.6829	0.4423	(2, 2) Even mode	$\alpha = 7.1580$	0.2195
0.85	0.9010	0.3271	0.8198	0.4819	(2, 2) Even mode	$\alpha = 7.0371$	0.2232
0.90	0.8641	0.2831	1.0198	0.5257	(2, 2) Even mode	$\alpha = 6.9368$	0.2264
0.95	0.8249	0.2119	1.3693	0.5743	(2, 2) Even mode	$\alpha = 6.8392$	0.2297

(continued)

Table. 4.1 (continued)

D_2/D_1	Pressure node of (2, 1) even mode $\left\{ \begin{array}{l} \xi_{(2,1)e} = 0, \\ y_{(2,1)e} = 0 \end{array} \right\}$		Pressure node of (0, 2) even mode $\{x_{(0,2)e} = 0\}$		Non-dimensional frequency range of broadband attenuation		Maximum ratio of the chamber length to the major-axis: $\frac{L}{D_1} \le \frac{\pi}{2\alpha}$
	$\eta_{(2,1)e}$	$\frac{x_{(2,1)e}}{D_1/2}$	$\xi_{(0,2)e}$	$\frac{y_{(0,2)e}}{D_1/2}$	Up to	$0 \le k_0 \frac{D_1}{2} \le \alpha$	
0.96	**0.8169**	**0.1917**	**1.4822**	**0.5846**	**(0, 3)** Even mode	$\alpha = \mathbf{7.1940}$	**0.2184**
0.97	0.8090	0.1678	1.6277	0.5951	(0, 3) Even mode	$\alpha = 7.1426$	0.2199
0.98	0.8011	0.1385	1.8323	0.6057	(0, 3) Even mode	$\alpha = 7.0955$	0.2214
0.99	0.7932	0.0990	2.1810	0.6166	(0, 3) Even mode	$\alpha = 7.0531$	0.2227
0.999	0.7862	0.0316	3.3344	0.6265	(0, 3) Even mode	$\alpha = 7.0191$	0.2238
1 (circular chamber)	$\frac{\delta_{(2,0)}}{D_0/2} = 0$		$\frac{\delta_{(0,1)}}{D_0/2} = 0.6276$		(0, 2) mode of 'circular' cylindrical chamber	$\alpha = 7.0156$	$\frac{L}{D_0} \le 0.2239$

The end-inlet port is located on the interfocal line $\delta_{E1} = x_{(2,1)e}$ (major-axis) and the end-outlet on the minor-axis at $\delta_{E2} = y_{(2,1)e}$

frequency range of interest, precisely, up to the cut-on frequency of the $(4, 1)e$ mode. The enhanced TL performance, characterized by multiple attenuation peaks, may be explained in light of the discussion in Sect. 3.4 on the influence of location of end ports on excitation or suppression of certain higher-order modes.

The variation of $[\mathbf{Z}]$ parameters with frequency (not included here) shows that at the resonance or cut-on frequencies of the $(1, 1)e$ and $(3, 1)e$ modes, the self-impedance parameter $Z_{11} \to \infty$ because the end port 1 is offset on the major-axis while the cross-impedance parameters Z_{21} and Z_{12}, and the self-impedance parameter Z_{22} are all finite because the end port 2 is offset on the minor-axis. Similarly, at the resonance frequencies of the $(1, 1)o$ and $(3, 1)o$ modes, $Z_{22} \to \infty$ while Z_{11}, Z_{21}, Z_{12} are all finite. In either case, one of the self-impedance parameters tends to infinity while others remain finite, which implies that one of the conditions set out in Sect. 3.6.1 for the occurrence of a peak in the TL spectrum is satisfied, thereby explaining a high attenuation at the above-mentioned resonance frequencies. Of particular importance, however, is the location of the end port 1 on the nodal hyperbola $\eta_{(2,1)e}$ due to which the contribution of the $(2, 1)e$ modal term in Z_{11} and Z_{21} (or Z_{12}) parameters is zero at all frequencies because $ce_2\left(\eta_{(2,1)e}, q_{2,1}\right) = 0$, see Denia et al. [23]. Therefore, these parameters are finite, while the parameter $Z_{22} \to \infty$ at the resonance frequency of the $(2, 1)e$ mode. This again explains the occurrence of an attenuation peak at the resonance frequency of this mode which significantly extends the range over which a broadband performance is obtained, thereby significantly enhancing the TL characteristics beyond this frequency [16]. Note that if the end port offset on the major-axis is not centered at the $(2, 1)e$ mode pressure node, a trough will be observed at the resonance frequency of the said mode, thereby leading to a breakdown of the broadband TL characteristics. The above arguments hold good regardless of the aspect-ratio of the short elliptical end-chamber.

On other hand, the location of the end port 2 on the nodal ellipse $\xi_{(0,2)e}$ results in a peak in the TL spectrum at the resonance frequency of the $(0, 2)e$ mode. This is due to zero contribution by the $(0, 2)e$ modal term in Z_{22} and Z_{21} (or Z_{12}) parameters at all frequencies because $Ce_0\left(\xi_{(0,2)e}, q_{0,2}\right) = 0$, while $Z_{11} \to \infty$ at the resonance frequency of this mode. However, for the particular $D_2/D_1 = 0.5$ considered, this peak is somewhat inconsequential due to the occurrence of a trough at a lower frequency, namely, at the resonance frequency of the $(4, 1)e$ mode leading to an early breakdown of the TL performance. Consequently, the broadband attenuation range is limited up to the non-dimensional frequency $\alpha = 6.4180$ as indicated in the Table 4.1. The trough may be explained by noting that all impedance parameters tend to infinity at resonance frequency of the $(4, 1)e$ mode implying that TL $\to 0$ as shown by Eq. (3.61). Further note that when the chamber length $L = 0.2448D_1$, the resonance frequencies of the first axial mode and the $(4, 1)e$ are coincident which implies that for a length greater than the above value, the trough due to the first axial mode will occur earlier than the $(4, 1)e$ mode trough, thereby limiting the broadband range and also the maximum L/D_1 ratio for a short chamber.

---

face or the side surface. An excellent agreement with experimental, finite-element or mode-matching methods throughout the frequency range of interest validated the piston-driven model.

Optimally locating the end ports, more precisely, the inlet port on the major-axis and the outlet port on the minor-axis at offset distances corresponding to the pressure nodes of (2, 1)e circumferential mode and (0, 2)e radial mode, respectively, with a view to obtain a maximum broadband attenuation range is henceforth referred to as the *double-tuning of short elliptical end-chamber mufflers*. It is similar to the practise of tuning the length of extended-inlet and outlet ducts of a straight-through concentric muffler configuration through appropriate end-correction [24]. Later in this chapter, we will also analyze configurations with a somewhat different arrangement of ports that can also deliver the same broadband frequency range provided that the eccentricity is either large, i.e., small aspect-ratio or it is nearly zero, i.e., sections tending to a perfect circle. The term double-tuning also applies to such elliptical or nearly circular chamber configurations.

4.2.1.1   Transition from a High-to-Zero Eccentricity Elliptical Section: Parametric Investigation

A parametric investigation is carried out to document and understand the changes in broadband acoustic attenuation spectrum as a highly eccentric elliptical section gradually transitions to a perfectly circular section (zero eccentricity) for double-tuned short end-chambers, i.e., a configuration with inlet and outlet ports offset on the major-axis and minor-axis, at the pressure nodes of the (2, 1)e and (0, 2)e modes, respectively. Table 4.1 lists the pressure nodal hyperbola $\eta_{(2,1)e}$ and ellipse $\xi_{(0,2)e}$ of the (2, 1)e mode and (0, 2)o modes, respectively, along with the corresponding non-dimensional offset distances of the end ports for a range of aspect-ratio commonly used for commercial silencers. Using data in Table 4.1 as well as the resonance frequency tables in Chap. 2, the TL performance of short flow-reversal and straight-through elliptical mufflers is computed, and the results presented in Fig. 4.3c up to Fig. 4.8 are systematically analyzed.

Figure 4.3c shows that for an elliptical section with $D_2/D_1 = 0.55$, the attenuation peak at the resonance frequency of the (0, 2)e mode occurs slightly before the trough near the resonance frequency of the (4,1)e mode. Although this enhances the TL spectrum in the region just before the (4,1)e mode trough, the broadband attenuation range itself remains unaltered and is given by $\alpha = \left(k_0 \frac{D_1}{2}\right)_{(4,1)e} = 6.3593$. Note that for elliptical chambers with $D_2/D_1 < 0.5$, the non-dimensional broadband attenuation range is also given by the resonance frequency of the (4,1)e circumferential mode as it is significantly smaller than its (0,2)e mode, i.e., the first radial mode counterpart—the reader is referred to Tables 2.8–2.10. It is worth mentioning that for $D_2/D_1 = 0.5231$, the resonance frequencies of the (4, 1)e and (0, 2)e modes are equal, and therefore, it is anticipated that the peak at (0, 2)e mode resonance in the TL graph would nullify the trough at the resonance frequency of the (4, 1)e mode which can possibly widen the broadband attenuation range. In this case, although a peak occurred at the (0, 2)e mode, unfortunately, it did not effectively lift or nullify the trough due to the (4, 1)e mode. As a result, the TL graph for

$D_2/D_1 = 0.5231$ (not shown here) exhibited nearly the same broadband range as observed for $D_2/D_1 = 0.55$, see Table 4.1.

Parts (a) and (b) of Figs. 4.4 and 4.5a show the TL graphs of double-tuned short end-chambers having $D_2/D_1 = 0.6, 0.65, 0.7$, respectively. In these results, although the trough or dip at the resonance frequency of the $(4, 1)e$ mode limits the broadband attenuation range, the TL produced at the trough increases with aspect-ratio. This is explained by the occurrence of a peak at the $(1, 2)e$ mode resonance which approaches the $(4, 1)e$ mode trough as the aspect-ratio increases from 0.6 to 0.7, refer to the resonance frequencies of these modes shown in Tables 2.13–2.15.

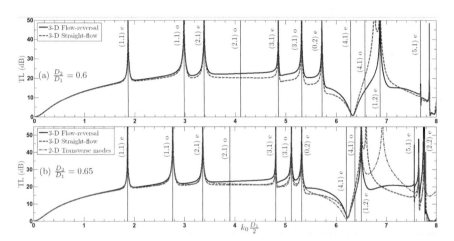

Fig. 4.4 TL performance of short flow-reversal and straight-flow elliptical end-chamber muffler configurations having **a** $D_2/D_1 = 0.6$ and **b** $D_2/D_1 = 0.65$

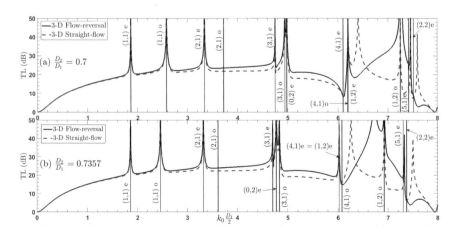

Fig. 4.5 TL performance of short flow-reversal and straight-flow elliptical end-chamber muffler configurations having **a** $D_2/D_1 = 0.7$ and **b** $D_2/D_1 = 0.7357$

Note that regardless of the aspect-ratio, a double-tuned short end-chamber configuration will always exhibit an attenuation peak at the $(1, 2)e$ mode resonance which becomes important for $D_2/D_1 = 0.7$ and beyond due to its ability to cancel the $(4, 1)e$ mode trough, thereby tending to increase the broadband range. Figure 4.5a shows a partial cancelation of the $(4, 1)e$ mode trough indicated by nearly 8 dB attenuation at the said frequency, beyond which a good attenuation performance is observed up to a frequency slightly less than the $(2, 2)e$ mode resonance. It is worth mentioning here that for $D_2/D_1 = 0.55$ and greater, the difference between the resonance frequencies of $(0, 2)e$ and $(4, 1)e$ modes progressively increases, i.e., the peak at the $(0, 2)e$ mode occurs before the $(4, 1)e$ mode trough. Therefore, offsetting the end-outlet port on the pressure nodal ellipse of the $(0, 2)e$ mode on minor-axis, i.e., double-tuning the short end-chamber is important for $D_2/D_1 \geq 0.55$ insofar as optimizing the broadband attenuation range is concerned.

At $D_2/D_1 = 0.7357$, the resonance frequencies of the $(1, 2)e$ and $(4, 1)e$ modes are identical. Figure 4.5b shows that this results in a complete cancelation of the $(4, 1)e$ mode trough by the $(1, 2)e$ mode resonance peak effectively increasing the broadband attenuation range up to a frequency slightly greater than the cut-on frequency of the $(2, 2)e$ mode given by $\alpha = 7.3649$ indicated in Table 4.1. (The local minima in the TL spectrum following the $(1, 2)e$ mode peak is approximately equal to 15 dB and occurs at the $(4, 1)o$ mode resonance.) Note that at the $(2, 2)e$ mode resonance, all $[\mathbf{Z}]$ matrix parameters tend to infinity in accordance with the discussion in Sect. 3.4 which implies that TL $\rightarrow$ 0, thereby explaining the breakdown in the broadband range in the vicinity of the $(2, 2)e$ mode resonance.

Figure 4.6a and b shows that the attenuation peak at the $(1, 2)e$ mode resonance is able to influence and significantly lift the $(4, 1)e$ trough for short end-chambers with $D_2/D_1 = 0.75$ notwithstanding the fact that the $(1, 2)e$ mode resonance occurs slightly before that of the $(4, 1)e$ mode. In fact, the attenuation produced at the dip frequency which immediately follows the $(4, 1)e$ mode resonance in Fig. 4.6 is 12 dB resulting in a broadband attenuation up to the $(2, 2)e$ mode resonance. Furthermore, Fig. 4.6 shows that in general, the 3-D semi-analytical uniform piston-driven model and 3-D FEA predictions are in a good agreement which again corroborates the former approach.

It is important to note that for elliptical sections with $D_2/D_1 > 0.7357$, the $(1, 2)e$ mode cut-on frequency occurs before the $(4, 1)e$ mode cut-on; however, their difference continues to be small, i.e., the resonance frequencies of the $(1, 2)e$ and $(4, 1)e$ modes are close, refer Tables 2.16–2.25. Consequently, the $(1, 2)e$ mode peak is able to lift the $(4, 1)e$ mode trough as may be observed from Figs. 4.6c and 4.7a and b which present the TL graph of double-tuned short end-chambers having $D_2/D_1 = 0.8, 0.85$ and 0.9, respectively. To this end, note that the attenuation produced at the dip occurring at $(4, 1)e$ mode resonance is approximately 8 dB, 14 dB and 20 dB, respectively, resulting in a broadband attenuation up to the $(2, 2)e$ mode resonance frequency indicated in Table 4.1. Based on these results, it is evident that for $D_2/D_1 \approx 0.7$ and greater, the $(1, 2)e$ mode peak is able to bring about a significant improvement in the TL graphs of double-tuned short chambers at higher frequencies.

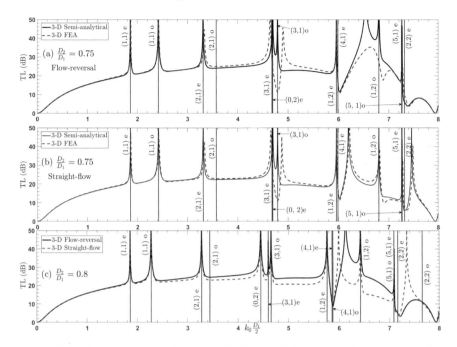

Fig. 4.6 TL performance of short **a** flow-reversal and **b** straight-flow elliptical end-chamber configurations having $D_2/D_1 = 0.75$. The 3-D semi-analytical approach is corroborated by comparing against the 3-D FEA results. **c** TL performance of short flow-reversal and straight-flow elliptical end-chamber configurations having $D_2/D_1 = 0.8$

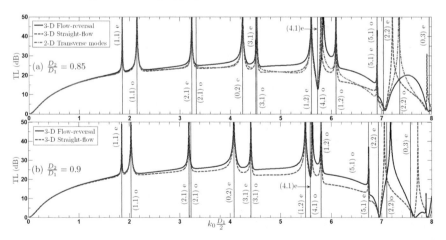

Fig. 4.7 TL performance of short flow-reversal and straight-flow elliptical end-chamber muffler configurations having **a** $D_2/D_1 = 0.85$ and **b** $D_2/D_1 = 0.9$

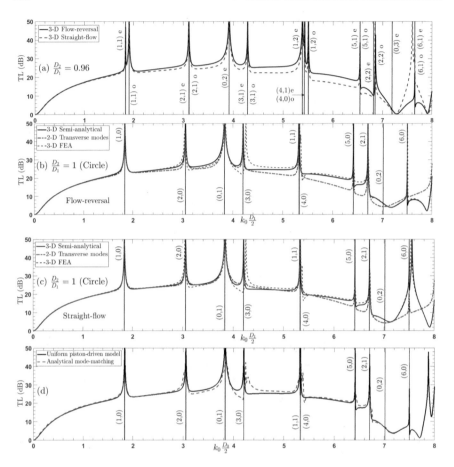

Fig. 4.8 **a** TL performance of short flow-reversal and straight-flow elliptical end-chamber configurations having $D_2/D_1 = 0.96$. TL performance of short (**b**) flow-reversal and (**c**) straight-flow
circular end-chamber configurations with end-centered inlet and end-outlet offset at $\delta_{E2} = \delta_{(0,1)} =
0.6276R_0$. The 3-D semi-analytical uniform piston-driven model is corroborated by comparing
against the 3-D FEA results. **d** Comparison of TL performance of a double-tuned short circular
flow-reversal muffler computed using the analytical mode-matching (Selamet and Ji, *J. Acoust.
Soc. Am.* 1998) and the uniform piston-driven model

For elliptical sections having $D_2/D_1 \geq 0.96$, the circumferential and cross-modes
of even and odd types tend to coalesce, and the TL graph approaches that of a
double-tuned short circular chamber with an end-centered inlet and optimally located
end-offset outlet [14]. This is demonstrated in Fig. 4.8a wherein it is observed that
attenuation peaks at even and odd modes almost coalesce. In particular, the trough
or dip occurring in the vicinity of the $(4, 1)e$ mode is completely raised, while an
additional peak is observed near the $(2, 2)e$ mode resonance extending the broadband
attenuation range up to the cut-on frequency of $(0, 3)e$ or the second radial mode

of the elliptical section. Similar results were observed for elliptical chambers with aspect-ratio tending to unity (circle) as indicated in Table 4.1.

Figures 4.8b and c show the TL performance of flow-reversal and straight-through double-tuned short circular muffler configurations, respectively, with an end-centered inlet, i.e., $\delta_{E1} = 0$ and end-outlet offset at a distance $\delta_{E2} = \delta_{(0,1)} = 0.6276R_0$ which is located on the pressure nodal circle of the $(0, 1)$ radial mode because $J_0\left(\alpha_{01}\frac{\delta_{(0,1)}}{R_0}\right) = J_0(2.4048) = 0$, see Ref. [14]. The predictions based on the uniform piston-driven model match reasonably well with the 3-D FEA results throughout the frequency range of interest. The TL graphs exhibit attenuation peaks at the resonance frequency of all circumferential and cross-modes, i.e., (m, n) modes where $m = 1, 2, 3, \ldots$ and $n = 0, 1, 2, \ldots$, and additionally at the resonance frequency of the $(0, 1)$ radial mode leading to a broadband attenuation up to the cut-on frequency of the $(0, 2)$ radial mode. These characteristics may be explained on the basis of the Green's function for the circular chamber shown in Eq. (3.13). It is observed that for the end-centered inlet port 1, all circumferential and cross-modes have a zero contribution in Z_{11}, Z_{21} and Z_{12} parameters for all frequencies while $Z_{22} \to \infty$ at resonance frequencies of these modes. On other hand, due to the offset location of the outlet port 2, the Z_{22}, Z_{21} and Z_{12} parameters are finite at the resonance frequency of the $(0, 1)$ radial mode while $Z_{11} \to \infty$ at this frequency. Furthermore, all impedance parameters tend to infinity at the $(0, 2)$ mode resonance. The above discussion explains the occurrence of peaks at circumferential and cross-modes, and the first radial mode as well as the occurrence of a trough near the $(0, 2)$ mode in Fig. 4.8b and c. Note that offsetting the outlet on the pressure nodal circle of the $(0, 1)$ radial mode significantly enhances the TL performance in the frequency band between the $(0, 1)$ and $(0, 2)$ mode resonances, much in the same manner as offsetting the outlet on the minor-axis at the pressure nodal ellipse of the $(0, 2)e$ radial mode extends the broadband range of the elliptical chamber up to the $(4, 1)e$, $(2, 2)e$ or $(0, 3)e$ modes depending on the aspect-ratio. In fact, if the outlet port center is located at an offset different from $\delta_{E2} = 0.6276R_0$, a trough will be observed at the $(0, 1)$ mode resonance frequency, thereby leading to the collapse of the broadband attenuation.

To further corroborate the mathematical model used here, the results are compared with the analytical mode-matching (AMM) results for a double-tuned short circular flow-reversing chamber configuration [14] in Fig. 4.8d. The chamber and port dimensions are given by $D_0 = 153.18$ mm, $L/D_0 = 0.205$, $d_0 = 24.3$ mm with the inlet port centered and the outlet port offset on the pressure node of the $(0, 1)$ radial mode at $\delta = 48.07$ mm. Again, a broadband attenuation up to the $(0, 2)$ radial mode and an excellent overall agreement throughout the frequency range is observed between the simpler uniform piston-driven model and the more rigorous AMM approach. Comparisons presented in Fig. 4.8b–d certainly validate the present mathematical model; however, small deviations in the frequency band between the $(0, 1)$ and $(1, 1)$ modes, particularly in the vicinity of a few attenuation peaks, are worth noting. Similar minor deviations from 3-D FEA predictions are also observed in the mid-frequency range in parts (a) and (b) of Figs. 4.3 and 4.6. As discussed earlier, this is attributed to the modeling of ports as uniform oscillating pistons, and the sound pressures on the inlet and outlet are averaged over the cross-sectional area which imply that the inlet and outlet are excluded in the mathematical model. Therefore, for short chambers, the resulting TL predictions in the absence of these ducts are expected to exhibit some deviation with respect to the 3-D FEA and mode-matching approaches as well as with experimental results. Nevertheless, in an overall sense, the comparisons validate and confirm the accuracy of the uniform piston-driven model for short chambers.

In the foregoing analysis for short chambers having $L/D_1 = 0.2$, the axial modes do not influence the maximum attenuation range because the first axial resonance $(k_L L = \pi)$ occurred at a frequency greater than the trough frequency at which the broadband pattern breaks down, i.e., at $(4, 1)e$, or $(2, 2)e$ or $(0, 3)e$ mode resonance denoted by $\alpha_{(m,n)}$. Therefore, by imposing the condition $k_\alpha \leq k_L$, the largest L/D_1 ratio corresponding to a maximum broadband attenuation range is obtained which is given by

$$\frac{L}{D_1} < \frac{\pi}{2\alpha_{(m,n)}}, \qquad (4.3)$$

which is tabulated in the last column of Table 4.1 for a range of aspect-ratio. (Here, k_α and k_L denote the wave numbers corresponding to the transverse and the first axial resonance frequency, respectively.) It follows from Eq. (4.3) that if a greater frequency range corresponding to a higher-order transverse mode over which the broadband nature is desired, then the L/D_1 ratio must be small. The foregoing TL graphs demonstrate that in this case, a somewhat flat and low attenuation is observed. On the other hand, if the broadband range is restricted to the first transverse mode, i.e., the $(1, 1)e$ or $(1, 0)$ mode, then the maximum $L/D_1 \approx 0.84$ which significantly improves the performance in the low-frequency range, i.e., within the axial plane wave limit, and is characterized by an attenuation peak or a dome depending on whether a flow-reversal [11] or straight-through type chamber [25] is analyzed. However, this is achieved at the expense of an early collapse of the spectrum.

For a double-tuned short circular muffler, the effect of taking the largest L/D_1 ratio indicated in Table 4.1 on the broadband attenuation range is examined. To this end, we present Fig. 4.9a and b which show the TL performance of flow-reversal and straight-through configurations, respectively, for $L = 0.2239 D_0$ keeping the other parameters same. Note that for this L/D_1 ratio, the resonance frequency of the first axial mode is coincident with that of the $(0, 2)$ or the second radial mode. (The 3-D semi-analytical and 3-D FEA results were found to be in a good agreement, thereby again validating the former.) Figure 4.9a shows that for the flow-reversal configuration, the attenuation peak produced at this frequency is able to significantly extend the broadband attenuation range up to the $(0, 3)$ or the third radial mode! This interesting outcome is referred to as the length-tuning of short flow-reversal circular chambers. However, Fig. 4.9b shows that the chamber length-tuning does not extend the broadband attenuation range for a straight-through muffler configuration, rather, the attenuation is only locally enhanced around this frequency. Furthermore, it was

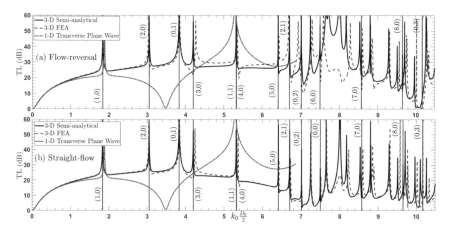

Fig. 4.9 TL performance of short (**a**) flow-reversal and (**b**) straight-flow circular end-chamber muffler configurations with a tuned axial length $L = 0.2239D_0$ having an end-centered inlet and end-outlet offset at $\delta_{E2} = \delta_{(0,1)} = 0.6276R_0$. The 3-D semi-analytical predictions are compared against the 3-D FEA results, and the results show that the transverse plane wave model deviates from the 3-D predictions just before the resonance frequency of $(1, 0)$ mode

found that for double-tuned short elliptical chambers of flow-reversal or straight-through type, taking the maximum L/D_1 ratio (length-tuning) unfortunately does not extend the broadband attenuation range, rather only a sharp peak similar to Fig. 4.9b was produced at the first axial mode resonance frequency.

A least-squares 10th degree polynomial (with coefficients correct to five significant digits) is fitted to the values of the pressure nodal hyperbola $\eta_{(2,1)e}$ of the $(2, 1)e$ mode and the non-dimensional offset distance $y_{(0,2)e}/(0.5D_1)$ corresponding to nodal ellipse of the $(0, 2)e$ mode for different aspect-ratio shown in Table 4.1 given by

$$\eta_{(2,1)e} = 0.928 - 0.11191\beta_1 - 0.02432\beta_1^2 + 0.0015273\beta_1^3 + 0.0037612\beta_1^4 + 0.001466\beta_1^5$$
$$- 9.7421 \times 10^{-5}\beta_1^6 - 0.0003483\beta_1^7 - 9.3584 \times 10^{-5}\beta_1^8 + 2.6073 \times 10^{-5}\beta_1^9 + 1.1681 \times 10^{-5}\beta_1^{10},$$

$$(4.4)$$

and

$$\frac{y_{(0,2)e}}{D_1/2} = 0.45908 + 0.13715\beta_2 + 0.023465\beta_2^2 + 0.0057184\beta_2^3 - 0.00035942\beta_2^4 - 0.0009785\beta_2^5$$
$$- 0.00045173\beta_2^6 - 9.0283 \times 10^{-5}\beta_2^7 + 7.3693 \times 10^{-5}\beta_2^8 + 6.8916 \times 10^{-5}\beta_2^9 + 1.6078 \times 10^{-5}\beta_2^{10},$$

$$(4.5)$$

where

$$\beta_1 = \frac{D_2/D_1 - 0.80993}{0.17395} \quad \text{and} \quad \beta_2 = \frac{D_2/D_1 - 0.82181}{0.17464}. \qquad (4.6a, b)$$

The interpolating polynomials given by Eqs. (4.4) and (4.5) can be used to accurately compute $\eta_{(2,1)e}$ and $y_{(0,2)e}/(0.5D_1)$ values, respectively, for any aspect-ratio between 0.35 and 1.0. Equation 4.5a is used to compute the optimal non-dimensional offset distance $x_{(2,1)e}/(D_1/2)$ of an end port along the major-axis for a given aspect-ratio, and this variation presented in Fig. 4.10a shows that for a circular chamber, the optimal offset is zero, i.e., the end port must be located at the center. This is because the nodal hyperbola $\eta_{(2,1)e}$ tends to the nodal diameter $\theta_{\text{nodal}} = \left\{ \frac{\pi}{4}, \frac{3\pi}{4} \right\}$ as the eccentricity goes to zero, refer to Fig. 2.7e. Figure 4.10a also shows the variation of the non-dimensional offset distance $y_{(0,2)e}/(0.5D_1)$ of the other end port along the minor-axis with aspect-ratio. It is observed that $y_{(0,2)e}/(0.5D_1)$ is maximum when $D_2/D_1 = 1$ and it is equal to $0.6276R_0$, i.e., the other port is offset on the pressure nodal circle of the (0, 1) radial mode. *Therefore, a short circular chamber with one of the end ports located at the center and other port located at optimal offset distance is the double-tuned counterpart of a highly eccentric short elliptical chamber with end-inlet and outlet ports located at optimal offset distance on the major- and minor-axis.* The muffler cross-sections presented in Fig. 4.10b schematically shows the transition of a double-tuned eccentric elliptical chamber to a double-tuned circular chamber.

Figure 4.11a and b shows the graphical variation of the broadband attenuation range α and the corresponding maximum L/D_1 ratio values of the double-tuned short end-chamber with its aspect-ratio D_2/D_1 presented in Table 4.1. Notice the jump or discontinuity in the graphs at certain aspect-ratio.

4.2.1.2 Acoustical Equivalence of Short Flow-Reversal and Straight-Through Mufflers: Two-Dimensional Analysis

In order to mathematically demonstrate the dominance of the higher-order transverse modes over the axial modes, a model based on the 2-D transverse modes is considered by altogether ignoring all those modes that have an axial dependence, i.e., we set $P = 0$ in the Green's function response given by Eq. (3.6) and the associated Eqs. (3.7)–(3.12). Therefore, the variation of the acoustic field along the axial direction is completely ignored owing to the short chamber length. In essence, the 2-D transverse model proposed here is the same as developed in previous papers for thin or short muffler elements of uniform rectangular or circular cross-section [17]. It delivers an accurate prediction throughout the frequency range of interest with deviations expected as one approaches the resonance frequency of the first axial mode.

The 2-D transverse model is compared with the more accurate 3-D approach for different aspect-ratio; the comparison results are presented in Figs. 4.3a, b, 4.4b, 4.7a and 4.8b, c for $D_2/D_1 = \{0.5, 0.65, 0.85, 1\}$, respectively. Generally speaking, an excellent agreement is observed between the 2-D and 3-D models up to the (2, 1)e and (2, 0) modes for elliptical and circular mufflers, respectively. In fact, the TL graphs of straight-through and flow-reversal configurations predicted by the 3-D approach are nearly coincident up to the frequency range noted above, beyond which noticeable deviations begin to show up, and the flow-reversal type muffler produces marginally more attenuation as compared to its straight-through counterpart. The 2-D model tends to break down near the first trough in the TL graphs, and as noted earlier,

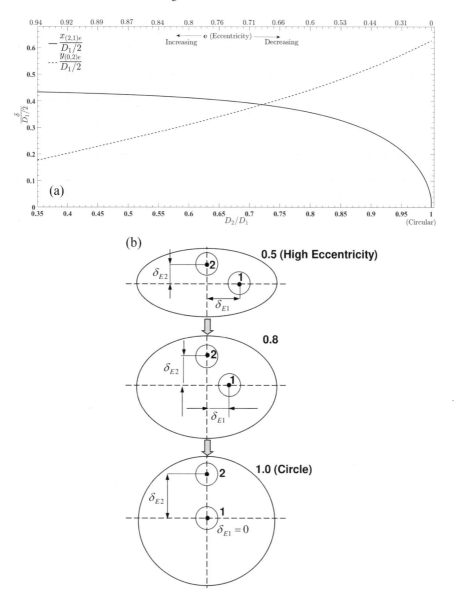

Fig. 4.10 a Variation of the optimal non-dimensional offset distance of end-inlet port $x_{(2, 1)e}/(D_1/2)$ located on the major-axis and end-outlet port $y_{(0, 2)e}/(D_1/2)$ located on the minor-axis, with aspect-ratio, **b** A schematic illustrating the gradual transition of a double-tuned highly eccentric elliptical chamber to a double-tuned circular chamber where the ports are located on the end face. The numbers indicate the aspect-ratio

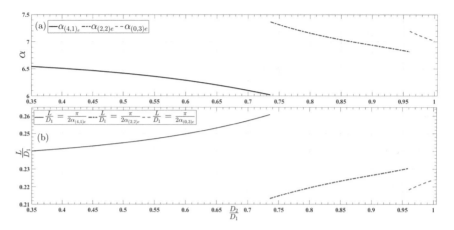

Fig. 4.11 Variation of the **a** non-dimensional frequency range of broadband attenuation α and **b** maximum L/D_1 ratio with aspect-ratio

this is expected due to the increased influence of the axial modes. Nevertheless, this comparison suggests the acoustical equivalence of short flow-reversal and straight-through configurations in the low-frequency range, thereby also demonstrating that the 2-D transverse model does not distinguish between the relative axial locations of the end ports. In fact, the same conclusion was arrived at in a previous paper [19] wherein short elliptical and circular end-chambers were analyzed using the 1-D transverse plane wave model using the Matrizant approach.

The 2-D nature may also be appreciated by showing the variation of the acoustic field inside short chambers at discrete frequencies. To this end, an analytical expression for the acoustic pressure field at a generic point (ξ, η, z) inside a flow-reversal type short elliptical muffler chamber subject to uniform piston excitation $u_1 = U_0$ at the inlet port 1 and anechoic condition/termination ($A_2 = 0$) at the outlet port 2 is considered below [11].

$$
\frac{p(\xi, \eta, z)}{\rho_0 c_0 U_0}
$$

$$
= jk_0 \left\{
\begin{array}{l}
\displaystyle\sum_{p=0,1,2,\dots}^{\infty} \sum_{m=0,1,2,\dots}^{\infty} \sum_{n=1,2,3,\dots}^{\infty} \frac{Ce_m(\xi, q_{m,n})ce_m(\eta, q_{m,n})\cos\left(\dfrac{P\pi z}{L}\right)\left\{\Delta_1^e - \left(\dfrac{S_{E2E1}}{1 - S_{E1E1}}\right)\Delta_2^e\right\}}{\left\{\left(\dfrac{P\pi}{L}\right)^2 + \dfrac{4q_{m,n}}{h^2} - k_0^2\right\}N_{m,n,P}} + \\[3ex]
\displaystyle\sum_{p=0,1,2,\dots}^{\infty} \sum_{m=1,2,\dots}^{\infty} \sum_{n=1,2,3,\dots}^{\infty} \frac{Se_m(\xi, \bar{q}_{m,n})se_m(\eta, \bar{q}_{m,n})\cos\left(\dfrac{P\pi z}{L}\right)\left\{\Delta_1^o - \left(\dfrac{S_{E2E1}}{1 - S_{E1E1}}\right)\Delta_2^0\right\}}{\left\{\left(\dfrac{P\pi}{L}\right)^2 + \dfrac{4\bar{q}_{m,n}}{h^2} - k_0^2\right\}\bar{N}_{m,n,P}}
\end{array}
\right\},
$$

$$(4.7)$$

where

$$\Delta_1^e = \iint\limits_{S_{E1}} \mathrm{Ce}_m\big(\xi_{E1}, q_{m,n}\big)\mathrm{ce}_m\big(\eta_{E1}, q_{m,n}\big)h_\xi h_\eta \mathrm{d}\xi_{E1}\mathrm{d}\eta_{E1},$$

$$\Delta_1^o = \iint\limits_{S_{E1}} \mathrm{Se}_m\big(\xi_{E1}, \overline{q}_{m,n}\big)\mathrm{se}_m\big(\eta_{E1}, \overline{q}_{m,n}\big)h_\xi h_\eta \mathrm{d}\xi_{E1}\mathrm{d}\eta_{E1}. \qquad (4.8a, b)$$

and

$$\Delta_2^e = \iint\limits_{S_{E2}} \mathrm{Ce}_m\big(\xi_{E2}, q_{m,n}\big)\mathrm{ce}_m\big(\eta_{E2}, q_{m,n}\big)h_\xi h_\eta \mathrm{d}\xi_{E2}\mathrm{d}\eta_{E2},$$

$$\Delta_2^o = \iint\limits_{S_{E2}} \mathrm{Se}_m\big(\xi_{E2}, \overline{q}_{m,n}\big)\mathrm{se}_m\big(\eta_{E2}, \overline{q}_{m,n}\big)h_\xi h_\eta \mathrm{d}\xi_{E2}\mathrm{d}\eta_{E2}. \qquad (4.9a, b)$$

Assuming plane wave propagation immediately from the port-chamber interface, the acoustic pressure field variation along the inlet port is calculated from the following relation between the progressive-wave amplitudes A_1, B_1 and the piston velocity amplitude U_0

$$\begin{Bmatrix} A_1 \\ B_1 \end{Bmatrix} = \rho_0 c_0 U_0 \begin{Bmatrix} \frac{1}{1-S_{11}} \\ \frac{S_{11}}{1-S_{11}} \end{Bmatrix}. \qquad (4.10)$$

The acoustic pressure field variation along the outlet port can similarly be worked out from the relation between the transmitted wave amplitude B_2 and piston velocity given by

$$B_2 = \rho_0 c_0 U_0 \frac{S_{21}}{1 - S_{11}}. \qquad (4.11)$$

Figure 4.12 shows the X–Z plane passing through the major-axis and the inlet port diameter over which the acoustic pressure field variation is analyzed inside a double-tuned flow-reversal type elliptical muffler configuration. As observed, the inlet and outlet ports are offset on the major-axis and minor-axis, respectively. Figure 4.13a–c shows the axial variation of the non-dimensional impedance $\log_{10}|p(\xi, \eta, z)/\rho_0 c_0 U_0|$ within the short elliptical chamber with $D_2/D_1 = 0.5$ and the inlet port 1 at non-dimensional frequency (a) $0.5k_0 D_1 = 1$, (b) $0.5k_0 D_1 = 1.8736$, i.e., the $(1, 1)e$ mode resonance, and (c) $0.5k_0 D_1 = 2.5$, while Fig. 4.14a–c shows the variation at non-dimensional frequency (d) $0.5k_0 D_1 = 4$, (e) $0.5k_0 D_1 = 6.4180$, i.e., the $(4, 1)e$ mode resonance and (f) $0.5k_0 D_1 = 7$. The arrow shows the piston velocity direction (considered positive looking into the system). Figure 4.13a, b shows that the acoustic pressure field does not vary along the axial direction at low frequencies, while only a marginal variation is observed at slightly higher frequency. These results demonstrate the 2-D nature of the acoustic pressure field within the chamber at these

Fig. 4.12 X–Z plane over
which axial variation of the
non-dimensional impedance
within the short elliptical
flow-reversal end-chamber
and the inlet port 1 is plotted

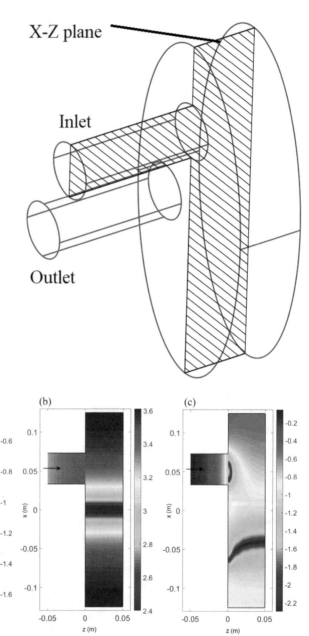

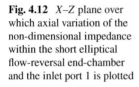

Fig. 4.13 Axial variation of the non-dimensional impedance $\log_{10}|p(\xi, \eta, z)/\rho_0 c_0 U_0|$ over the X–Z plane within the short elliptical flow-reversal end-chamber having $D_2/D_1 = 0.5$ at non-dimensional frequency: **a** $0.5k_0 D_1 = 1$, **b** $0.5k_0 D_1 = 1.8736$, and **c** $0.5k_0 D_1 = 2.5$

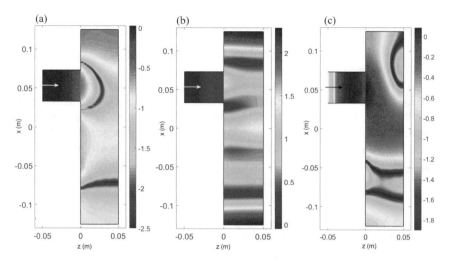

Fig. 4.14 Axial variation of the non-dimensional impedance $\log_{10}|p(\xi, \eta, z)/\rho_0 c_0 U_0|$ over the X–Z plane within the short elliptical flow-reversal end-chamber having $D_2/D_1 = 0.5$ at non-dimensional frequency: **a** $0.5k_0 D_1 = 4$, **b** $0.5k_0 D_1 = 6.4180$, and **c** $0.5k_0 D_1 = 7$

frequencies, thereby corroborating the 2-D transverse model. At higher frequencies, the axial variation becomes more readily noticeable as may be observed from the pressure contours shown in Fig. 4.14a–c. In particular, Fig. 4.14c shows a significant variation along both X and Z directions, thereby suggesting a breakdown of the 2-D transverse mode model, and existence of a 3-D acoustic pressure field at such high frequencies. Also note the discontinuity in the acoustic pressure field at the port-chamber interface which is due to the uniform piston-driven model approximation (refer to the discussion in Sect. 3.2.2.1).

4.2.1.3 Short Elliptical Chambers of Equal Cross-Sectional Area: Summarized Results

For a family of ellipses of a constant cross-sectional area [26], the maximum broadband attenuation range of double-tuned straight-flow or flow-reversal configurations is significantly different from their counterpart members belonging to the family of ellipses with a constant major-axis discussed in the foregoing sections. The objective here is to discuss the changes in the broadband attenuation spectrum as a perfectly circular cross-section is deformed into an elliptical cross-section with a high eccentricity while keeping the cross-sectional area constant. To this end, we consider a circular chamber of diameter D_{eq} which say, is deformed into an elliptical chamber with a given aspect-ratio D_2/D_1. On using Eq. (2.70), one obtains the major-axis

$D_1 = D_{eq}\left(\frac{D_2}{D_1}\right)^{-\frac{1}{2}}$, and for such an ellipse, the maximum non-dimensional broad-band attenuation range can be found using Eq. (2.71) where $\alpha_{m,n}^{elliptical}$ can be found from Table 4.1 or from Fig. 4.11a.

Table 4.2 shows the non-dimensional frequency range of broadband attenuation for a family of ellipses with a constant cross-sectional area, with aspect-ratio varying from 0.35 to 1. It is observed that maximum frequency range of broadband attenuation increases with aspect-ratio which is unlike Table 4.1 or the variation shown in Fig. 4.11a. Although the TL graphs are not included here for brevity, the following results are discussed. A highly eccentric ellipse has a significantly smaller attenuation range as compared to a circular chamber, thereby signifying that the breakdown

Table. 4.2 Non-dimensional frequency range of broadband attenuation of double-tuned short elliptical chambers having the same cross-sectional area

$\dfrac{D_2}{D_1}$	Non-dimensional frequency Range of broadband attenuation $\left(k_0\frac{D_{eq}}{2}\right)_{m,n} = \sqrt[4]{1-e^2} \times \alpha_{m,n}^{elliptical} = \sqrt{\frac{D_2}{D_1}} \times \alpha_{m,n}^{elliptical}$	
	Up to	$0 \le \left(k_0\frac{D_{eq}}{2}\right)_{m,n} \le \alpha$
0.35	(4, 1) even mode	$\alpha = 3.8704$
0.40	(4, 1) even mode	$\alpha = 4.1158$
0.45	(4, 1) even mode	$\alpha = 4.3381$
0.50	(4, 1) even mode	$\alpha = 4.5382$
0.5231	(4, 1) even mode	$\alpha = 4.6155$
0.55	(4, 1) even mode	$\alpha = 4.7162$
0.60	(4, 1) even mode	$\alpha = 4.8711$
0.65	(4, 1) even mode	$\alpha = 5.0012$
0.70	(4, 1) even mode	$\alpha = 5.1056$
0.7357	**(2, 2)** even mode	$\alpha = 6.3171$
0.75	(2, 2) even mode	$\alpha = 6.3329$
0.80	(2, 2) even mode	$\alpha = 6.4023$
0.85	(2, 2) even mode	$\alpha = 6.4879$
0.90	(2, 2) even mode	$\alpha = 6.5808$
0.95	(2, 2) even mode	$\alpha = 6.6660$
0.96	**(0, 3)** even mode	$\alpha = 7.0489$
0.97	(0, 3) even mode	$\alpha = 7.0346$
0.98	(0, 3) even mode	$\alpha = 7.0242$
0.99	(0, 3) even mode	$\alpha = 7.0177$
0.999	(0, 3) even mode	$\alpha = 7.0156$
1 (circular chamber)	**(0, 2)** mode of 'circular' cylindrical chamber	$\alpha = 7.0156$

or occurrence of a trough much earlier in the spectrum—this was expected because the major-axis is larger than the diameter. Nevertheless, up to the $(1, 1)e$ mode resonance, the TL graphs for this family were found to be nearly identical because in the low-frequency range, the attenuation depends only on the area expansion ratio [27] which is constant by definition. Furthermore, a rather high attenuation characterized by peak(s) was observed over the frequency range just before the breakdown.

4.2.2 Inlet on the Major-Axis and Outlet Port Centered

Short elliptical muffler configurations having end-inlet port 1 centered on the intersection of the pressure nodal hyperbola $\eta_{(2,1)e}$ of the $(2, 1)e$ mode and the major-axis, and the end-outlet port 2 located at the center of ellipse are analyzed in the section. In other words, the offset location of end-inlet port 1 along the major-axis is given by Eq. (4.1) while for the end-outlet port 2, $\delta_{E2} = 0$. The *configuration 1* in Fig. 4.2c shows the location of end ports described above.

Figures 4.15a, b and 4.16a, b show the TL performance of the aforementioned configuration 1 for aspect-ratio $D_2/D_1 = 0.35, 0.5, 0.65$, and 0.85, respectively, computed using the 3-D semi-analytical approach. The TL graphs for both flow-reversal and straight-through type mufflers are presented except in Fig. 4.16b where only the latter is analyzed. First note that for the port located at the ellipse center, only the even-even modes are excited while for the other port offset on the major-axis, only the even-even and even-odd modes are excited. Therefore, for the configurations

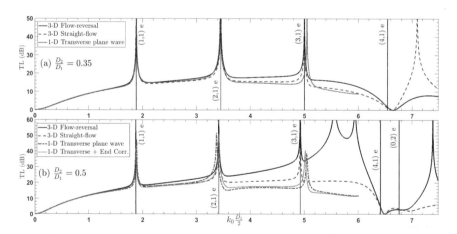

Fig. 4.15 TL performance of short flow-reversal and straight-flow elliptical end-chamber configurations having **a** $D_2/D_1 = 0.35$ and **b** $D_2/D_1 = 0.5$ with the end-inlet port located on the major-axis ($\delta_{E1} = x_{(2,1)e}$) and end-outlet port centered. The 1-D transverse plane wave model is shown to fail at the resonance frequency of $(3, 1)e$ mode in (**a**) and at $(2, 1)e$ mode in (**b**). By considering appropriate end-corrections in (**b**), the accuracy of the 1-D model was marginally improved

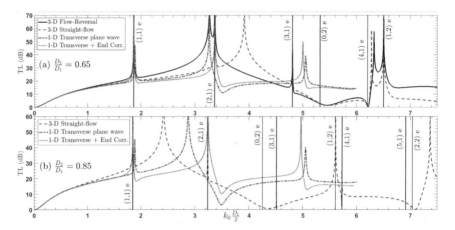

Fig. 4.16 TL performance of short elliptical end-chamber configurations having **a** $D_2/D_1 = 0.65$ and **b** $D_2/D_1 = 0.85$ with the end-inlet port located on the major-axis ($\delta_{E1} = x_{(2,1)e}$) and end-outlet port centered. By considering end-corrections, the accuracy of the 1-D transverse plane wave model was improved up to the $(2, 1)e$ and $(1, 1)e$ modes in **(a)** and **(b)**, respectively

1 and 4 shown in Fig. 4.2c, the odd modes do not participate or contribute to the modal solution, and thus, one needs to consider only the even modes in the acoustic pressure response solution shown in Eq. (3.6). Figures 4.15 and 4.16 show that at the resonance frequencies of the $(1, 1)e$ and $(3, 1)e$ modes, the self-impedance parameter $Z_{11} \rightarrow \infty$ because the end port 1 is offset on the major-axis, while the cross-impedance parameters Z_{21} and Z_{12}, and the self-impedance parameter Z_{22} are all finite because the end port 2 is located at the ellipse center. Similarly, at the resonance frequency of the $(2, 1)e$ mode, $Z_{22} \rightarrow \infty$ while Z_{11}, Z_{21}, Z_{12} are all finite because the end port 1 is centered on the pressure nodal hyperbola $\eta_{(2,1)e}$. In either case, one of the self-impedance parameters tends to infinity, while others remain finite at a given resonance frequency, which explains the occurrence of an attenuation peak in the TL spectrum. Further note that for configuration 1, the breakdown of the broadband attenuation will occur either at the resonance frequency of the $(4,1)e$ circumferential mode or the $(0, 2)e$ radial mode whichever occurs earlier because these modes cannot be suppressed by either ports.

Figure 4.15a shows that the breakdown of the broadband attenuation occurs at the resonance frequency of the $(4, 1)e$ mode because the $(0, 2)e$ mode resonance mode is significantly greater for $D_2/D_1 = 0.35$, refer to Table 2.8. In fact, the broadband but relatively flat attenuation characteristics for both flow-reversal and straight-through configurations in Fig. 4.15a resemble the TL graphs of their counterpart double-tuned configuration 2. A similar result is observed from Fig. 4.15b; here too, the broadband attenuation range is up to the $(4, 1)e$ mode resonance because the $(0, 2)e$ mode resonance occurs at a slightly greater frequency. However, a much greater attenuation is produced for the flow-reversal chamber and is characterized by occurrence of multiple peaks in the frequency range between the $(3, 1)e$ and $(4, 1)e$ mode resonances before the collapse at the $(4, 1)e$ mode. Therefore, up to $D_2/D_1 \approx 0.5$, the configurations 1 and 2 are equivalent insofar as double-tuning is concerned provided that the end port 1 is centered at the $\eta_{(2, 1)e}$ pressure nodal hyperbola on the major-axis.

With a further increase in aspect-ratio, the broadband attenuation characteristics deteriorate rather rapidly because for $D_2/D_1 > 0.5231$, the $(0, 2)e$ mode resonance frequency occurs before the $(4, 1)e$ mode resonance. This is demonstrated by the results shown in Fig. 4.16a and b wherein it is observed that the collapse of the broadband pattern occurs much earlier, precisely at $\alpha_{(0,2)e} = 5.3202$ and 4.2487, respectively. In fact, beyond the $(1, 1)e$ mode resonance frequency, the TL graphs in Fig. 4.16a, b resemble that of a side-branch resonator, and in an overall sense, the attenuation characteristics are significantly different from that of a double-tuned short elliptical configuration 1 analyzed in Sect. 4.2.1. In light of these results, it is concluded that only for highly eccentric elliptical chambers having $D_2/D_1 \approx 0.5$ or smaller, configurations 1 and 2 are equivalent insofar as double-tuning is concerned. This is most likely attributed to the shape of the even-even and even-odd circumferential modes for narrow or eccentric ellipses. The pressure nodal hyperbolas of these modes are oriented in a direction nearly parallel to the minor-axis, and a standing wave pattern is formed along the major-axis, see Fig. 4.17a–d which show the first four even circumferential modes for $D_2/D_1 = 0.35$, and refer back to Fig. 2.6a–c which show the first three even circumferential modes for $D_2/D_1 = 0.5$. Figure 4.18a–c shows the variation of normalized non-dimensional impedance $|p(\xi, \eta, z)/\rho_0 c_0 U_0|$ over the elliptical cross-section with $D_2/D_1 = 0.5$ at different frequencies. These results pertain to the muffler configuration 2 of the flow-reversal type analyzed in Fig. 4.15b and were computed by making using of Eq. (4.7) over the $\eta - \xi$ plane passing through $z = 0.5L$. It is evident from Fig. 4.18a–c that the wave propagation is predominantly along the major-axis, even up to the resonance frequencies of higher-order even circumferential modes.

For ellipses of a larger aspect-ratio, the nodal hyperbolas of the higher-order circumferential modes are no longer oriented perpendicular to the major-axis, rather, their curvilinearity is pronounced, see Fig. 4.19a–c and d–f which show the first three even circumferential modes for $D_2/D_1 = 0.65$ and 0.85, respectively. These mode shapes suggest that beyond the $(1, 1)e$ mode resonance frequency, wave propagation cannot be considered along the major-axis, rather, the acoustic pressure field for such short chambers has a two-dimensional nature. This is also confirmed from Fig. 4.18d–f which shows the variation of normalized non-dimensional impedance $|p(r, \theta, z)/\rho_0 c_0 U_0|$ over the circular cross-section at different frequencies for the double-tuned short circular muffler analyzed in Fig. 4.8b.

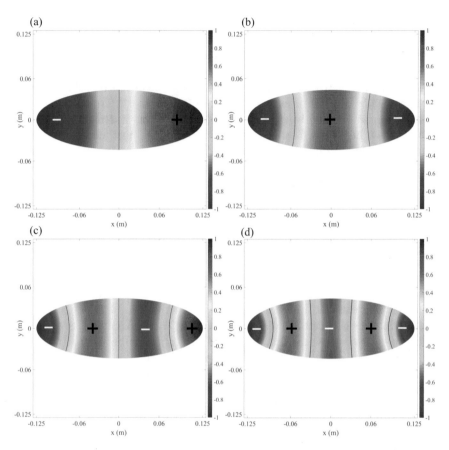

Fig. 4.17 Mode shapes of the first four even circumferential modes of the ellipse with $D_2/D_1 = 0.35$

4.2.2.1 The One-Dimensional Transverse Plane Wave Model

In order to further corroborate the above discussion, the TL performance for configuration 1 for different aspect-ratio using the 1-D transverse plane wave approach [18] is computed, and the results are compared with those obtained using the 3-D semi-analytical approach. In essence, the transverse plane wave considers propagation along the major-axis for short elliptical chambers and by its definition is valid only for configurations 1 and 4 in which the ports are located on the major-axis or at the ellipse center which allows only the even-even and even-odd modes to be excited. The Webster's horn equation [28] which governs wave propagation in a gradually varying area duct is used to model the 1-D transverse plane wave propagation and is given by

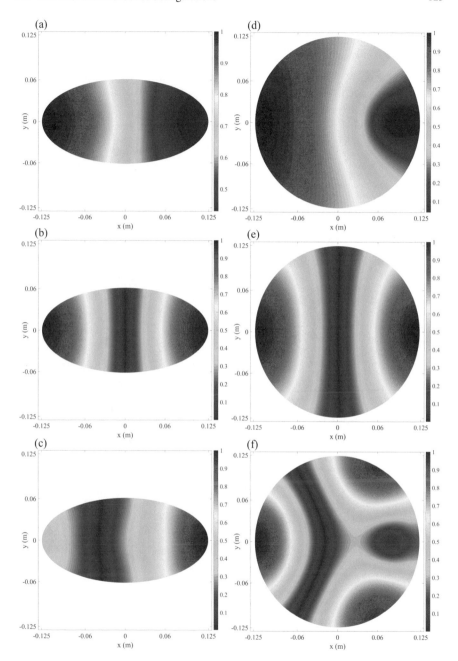

Fig. 4.18 Variation of the normalized non-dimensional impedance $|p(\xi, \eta, z)/\rho_0 c_0 U_0|$ evaluated at the axial distance $z = 0.5L$ over the cross-section of short elliptical chamber $(D_2/D_1 = 0.5)$ in parts (**a–c**) and over the cross-section of short circular chamber in parts (**d–f**). Parts **a** and **d** are evaluated at $0.5k_0D_1 = 1$, while parts **c** and **f** are evaluated at $0.5k_0D_1 = 2.5$. Parts **b** and **e** are evaluated at the resonance frequency of the $(1, 1)e$ and $(1, 0)$ modes, respectively

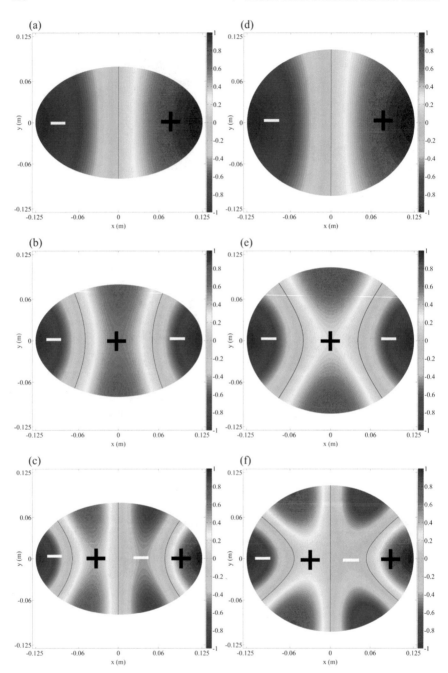

Fig. 4.19 a–c Mode shapes of the first three even circumferential modes of the ellipse having $D_2/D_1 = 0.65$. **d–f** Mode shapes of the first three even circumferential modes of the ellipse having $D_2/D_1 = 0.85$

$$\frac{d^2 p}{dx'^2} + \frac{1}{S(x')} \frac{dS(x')}{dx'} \frac{dp}{dx'} + k_0^2 p = 0, \tag{4.12}$$

where $p = p(x')$ denotes the 1-D acoustic pressure field, the coordinate x' is measured from the top of the ellipse (see Fig. 4.20a), and $S(x')$ denotes the cross-sectional area perpendicular to which the plane wave propagates along the direction of the major-axis. In the non-dimensional form, the cross-sectional area is given by

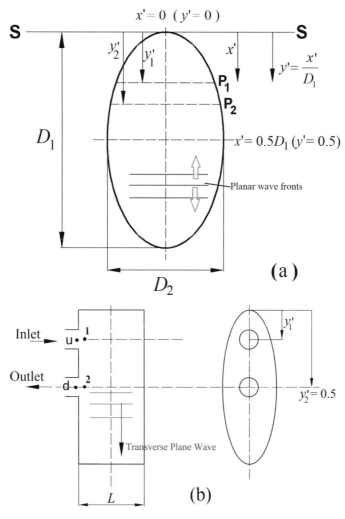

(a)

(b)

Fig. 4.20 Part **a**: An elliptical cross-section with plane wave propagation along the major-axis given by D_1. The distance x and the non-dimensional length y are measured from the top section S–S. Part **b**: A short elliptical end-chamber muffler with end-inlet offset on the major-axis and end-inlet port located at ellipse center

$$S(y') = 2D_2L\sqrt{y' - y'^2},\qquad(4.13a)$$

and

$$y' = x'/D_1,\qquad(4.13b)$$

denotes the non-dimensional coordinate also shown in Fig. 4.20a. Note that Eq. (4.12) implies a stationary medium, i.e., zero mean flow.

An analytical solution for Eq. (4.12) is expressed as two linearly independent solutions $F_1(y', \beta)$ and $F_2(y', \beta)$ obtained by the method of Frobenius and is given by

$$p(y', \beta) = A\, F_1(y', \beta) + B\, F_2(y', \beta),$$
$$F_1(y', \beta) = P_0(\beta) + P_1(\beta)y' + P_2(\beta)y'^2 + P_3(\beta)y'^3 + \cdots,$$
$$F_2(y', \beta) = Q_0(\beta)y'^{1/2} + Q_1(\beta)y'^{3/2} + Q_2(\beta)y'^{5/2} + Q_3(\beta)y'^{7/2} + \cdots.$$
$$(4.14a\text{--}c)$$

where $\beta = k_0 D_1$ is the non-dimensional frequency. The functions $F_1(y', \beta)$ and $F_2(y', \beta)$ were obtained by considerable computer algebra (on the symbolic package MAPLE [29]), and for the purpose of implementation, the series shown in Eqs. (4.14b, c) was truncated to the first 13 terms or sub-functions $P_i(\cdot)$ and $Q_i(\cdot)$ given by

$$P_0(\beta) = 1,$$
$$P_1(\beta) = 0,$$
$$P_2(\beta) = -\beta^2/3,$$
$$P_3(\beta) = -2\beta^2/45,$$
$$P_4(\beta) = -\beta^2/35 + \beta^4/42,$$
$$P_5(\beta) = -32\beta^2/1575 + 5\beta^4/1222,$$
$$P_6(\beta) = -32\beta^2/2079 + 7\beta^4/2673 - \beta^6/1386,$$
$$P_7(\beta) = -53\beta^2/4352 + 15\beta^4/7934 - 4\beta^6/29091,$$
$$P_8(\beta) = -64\beta^2/6435 + 23\beta^4/15733 - \beta^6/11397 + \beta^8/83160,$$
$$P_9(\beta) = -27\beta^2/3245 + 11\beta^4/9314 - 13\beta^6/203486 + \beta^8/412178,$$
$$P_{10}(\beta) = -24\beta^2/3383 + 5\beta^4/5084 - 2\beta^6/40033 + \beta^8/647316 - \beta^{10}/7900200,$$
$$P_{11}(\beta) = -31\beta^2/5047 + 7\beta^4/8359 - \beta^6/24496 + \beta^8/884188 - \beta^{10}/37757077,$$
$$P_{12}(\beta) = -25\beta^2/4642 + 5\beta^4/6893 - 3\beta^6/87305 + \beta^8/1122670$$
$$\qquad\qquad - \beta^{10}/59395622 + \beta^{12}/1090227600,\qquad(4.15)$$

and

$$Q_0(\beta) = 1,$$
$$Q_1(\beta) = 1/6,$$
$$Q_2(\beta) = 3/40 - \beta^2/5,$$
$$Q_3(\beta) = 5/112 - 5\beta^2/126,$$
$$Q_4(\beta) = 35/1152 - 71\beta^2/3240 + \beta^4/90,$$
$$Q_5(\beta) = 63/2816 - 44\beta^2/2927 + 13\beta^4/5527,$$
$$Q_6(\beta) = 94/5417 - 24\beta^2/2125 + 12\beta^4/8767 - \beta^6/3510,$$
$$Q_7(\beta) = 23/1647 - 10\beta^2/1119 + 9\beta^4/9274 - 4\beta^6/64063,$$
$$Q_8(\beta) = 30/2597 - 25\beta^2/3416 + 19\beta^4/25408 - \beta^6/26890 + \beta^8/238680,$$
$$Q_9(\beta) = 52/5327 - 27\beta^2/4394 + 7\beta^4/11583 - \beta^6/37359 + \beta^8/1065491,$$
$$Q_{10}(\beta) = 92/10965 - 6\beta^2/1141 + \beta^4/1984 - \beta^6/47878 + \beta^8/1766099$$
$$\qquad\qquad - \beta^{10}/25061400,$$
$$Q_{11}(\beta) = 153/20923 - 54\beta^2/11819 + 9\beta^4/20930 - \beta^6/58585 + \beta^8/2432504$$
$$\qquad\qquad - \beta^{10}/110189705,$$
$$Q_{12}(\beta) = 19/2947 - 16\beta^2/3981 + 7\beta^4/18756 - \beta^6/69538 + \beta^8/3093608$$
$$\qquad\qquad - \beta^{10}/181183963 + \beta^{12}/3759210000, \qquad\qquad (4.16)$$

The solution given by Eqs. (4.15) and (4.16) corresponds to the roots $m_r = 0$ and $1/2$, respectively, and the solution exhibits a good convergence over the half-domain $0 \leq y' \leq 0.5$ and up to the non-dimensional frequency up to $\beta = k_0 D_1 \approx 9$, i.e., $k_0 D_1/2 = 4.5$. For sound speed at room temperature and major-axis up to 250 mm, the maximum frequency range up to which a convergent solution can be obtained is approximately 1965 Hz which is sufficient as a significant portion of the exhaust noise of automotive mufflers is limited to only the first few harmonics of the firing frequency.

Figure 4.20b presents the front-view of a short elliptical chamber showing transverse plane wave propagation along its major-axis, as well as the cross-sectional view (same as configuration 1 in Fig. 4.2c) showing the location of the end-inlet port 1 at a certain offset distance y_1' and end-outlet port 2 at the center, i.e., at $y_2' = 0.5$. For this flow-reversal elliptical configuration, the detailed derivation of the $[\mathbf{T}_{ud}]$ matrix between the point u in the end-offset inlet port and the point d in the end-centered outlet port is presented in Ref. [18]. Here, we present only its simplified version given by

$$\left\{ \begin{matrix} p_u \\ v_u \end{matrix} \right\} = \begin{bmatrix} 1 & 0 \\ 1/Z_1 & 1 \end{bmatrix} \underbrace{\begin{bmatrix} T_{11} & T_{12} \\ T_{21} & T_{22} \end{bmatrix}_{y_1 y_2 = 0.5}}_{[\mathbf{T}_{y1y2=0.5}]} \begin{bmatrix} 1 & 0 \\ 1/Z_2 & 1 \end{bmatrix} \left\{ \begin{matrix} p_d \\ v_d \end{matrix} \right\}, \qquad (4.17)$$

$$[\mathbf{T}_{ud}]$$

where the shunt-impedances Z_1 and Z_2 are given by

$$Z_1 = j\beta Y_1 \frac{F_1(y_1')}{F_1'(y_1')}, \quad Y_1 = \frac{c_0}{S(y_1')}, \quad F_1'(y_1') = \left. \frac{d}{dy} F_1(y') \right|_{y=y_1}, \qquad (4.18a\text{--}c)$$

and

$$Z_2 = j\beta Y_m \frac{F_1(y_2' = 0.5)}{F_1'(y_2' = 0.5)}, \quad Y_m = \frac{c_0}{S(0.5)}, \quad F_1'(0.5) = \left. \frac{d}{dy} F_1(y') \right|_{y=0.5}. \\ \qquad\qquad (4.19a\text{--}c)$$

The analogous electro-acoustic circuit represented by Eq. (4.17) consists of shunt-impedances Z_1 and Z_2 pertaining to the side-branch cavities formed above the end-offset port 1 and below the end-centered port 2, respectively, while the transverse plane wave propagation between the ports shown in Fig. 4.20b is a distributed element [19, 27]. Furthermore, the transverse plane wave model inherently considers the short flow-reversal and straight-through type muffler configurations to be acoustically equivalent. Also note that due to acoustic reciprocity [30], the TL performance for the configuration shown in Fig. 4.20b will remain unaltered if the inlet and outlet locations are interchanged, i.e., the inlet is located at the ellipse (circle) center and outlet is offset on the major-axis under the assumption of zero mean flow.

Figures 4.15a, b and 4.16a, b compare the TL prediction obtained using the 1-D transverse plane wave approach with that computed using the more accurate 3-D semi-analytical uniform piston-driven model. Note that to compute the 1-D results, the location y_1' of the end-offset port 1 on the major-axis corresponds to the same offset $x_{(2,1)e}/(D_1/2)$ indicated in Table 4.1 for a given aspect-ratio, while the end port 2 is located at $y_2' = 0.5$ which coincides with the ellipse center. Figure 4.15a demonstrates a good agreement between the 1-D and 3-D approaches nearly up to (3, 1)e mode resonance for $D_2/D_1 = 0.35$ which can be explained on the basis of mode shapes shown in Fig. 4.17. Beyond the (3, 1)e mode resonance, significant deviations are observed, particularly from the flow-reversal muffler results, thereby leading to the breakdown of the 1-D model. Figure 4.15b shows that for $D_2/D_1 = 0.5$, the 1-D model breaks down slightly after the (2, 1)e mode resonance, while for $D_2/D_1 = 0.65$ and 0.85, the validity of the 1-D model is limited to the frequency range between the (1, 1)e and (2, 1)e mode resonance, and up to the (1, 1)e mode resonance, respectively. Indeed, for a short circular chamber, the transverse plane wave approach is less suited as may be appreciated by observing the significant deviation between the 1-D and 3-D predictions even before the onset of the (1, 0) mode in Fig. 4.9a, b. Note that the more accurate 3-D approach delivers the attenuation peak exactly at the $(1, 0)$ mode resonance while the 1-D model produces a significant deviation. This result, however, is not surprising and can readily be explained by observing the $(1, 0)$ mode shape shown in Fig. 2.7a as well as the acoustic field variation shown in Fig. 4.18d.

Resonance Frequencies and End-Correction

Notwithstanding the deviations at higher frequencies, the above discussion suggests that owing to its simplicity, the 1-D transverse plane wave model is a useful tool insofar as a quick analysis of short end-chamber mufflers is concerned, at least up to the low-frequency limit. While the numerical schemes such as the stepped-segmentation [31] and the Matrizant approach [19] were shown to yield identical results, the analytical Frobenius series solution offers an important additional advantage. It allows the use of 1-D transverse plane wave model for double-tuning the TL performance of highly eccentric elliptical chambers. This is explained as follows. The non-dimensional frequency at which the transverse plane wave troughs occur is found by numerically solving [18]

$$F_1(0.5)F_1'(0.5) = 0. \tag{4.20}$$

The first trough frequency is given by $k_0 D_1/2 = 1.8866$ obtained by solving $F_1(0.5) = 0$, while the resonance frequency of the cavity formed below the end-centered port 2 can be found by setting $Z_2 = 0$ which on using Eq. (4.19a), incidentally leads to the same equation $F_1(0.5) = 0$. This implies that the non-dimensional resonance frequency of the cavity is the same as the frequency of occurrence of the first transverse trough of the short chamber. Indeed this non-dimensional frequency $k_0 D_1/2 = 1.8866$ is found to be identical with the $(1, 1)e$ mode of the limiting case of the highly eccentric ribbon-like ellipse having $D_2/D_1 = 0.01$ (see Table 2.1), and remains quite close to this mode regardless of the aspect-ratio. This not only further confirms the validity of the 1-D transverse plane wave model up to this frequency, but also suggests that single-tuning can always be achieved regardless of the aspect-ratio by simply locating one of the end ports at the ellipse center because the attenuation peak at the resonance frequency of this cavity completely annihilates the first transverse trough.

Now, the second trough frequency is given by $k_0 D_1/2 = 3.4865$ which was obtained by solving $F_1'(0.5) = 0$. Therefore, in order to double-tune the short elliptical chamber based on the 1-D transverse plane wave model, the offset location y_1' of the end port 1 should be such that the resonance frequency of the shorter cavity formed above this port must be coincident with the second transverse trough frequency noted above. To this end, the shunt-impedance Z_1 is set to zero which on using Eq. (4.18a), yields

$$F_1(y_1') = 0. \tag{4.21}$$

Equation (4.21) is numerically solved for y_1' such that $\beta/2 = k_0 D_1/2 \approx 3.4865$. Using a root-bracketing iterative procedure, it was found that $y_1' = 0.2795$. Therefore, the optimal non-dimensional offset location of the end port 1 along the major-axis measured from the ellipse center is given by

$$\frac{x}{D_1/2} = 1 - 2y_1' = 0.4409. \tag{4.22}$$

The 1-D transverse plane wave model, therefore, suggests that regardless of its aspect-ratio, a short elliptical chamber muffler can be double-tuned by locating one of the end ports at the center and offsetting the other end port on the major-axis where the offset distance is given by Eq. (4.22). It also implies that if the offset distance of the end port 1 is significantly different than the one specified by Eq. (4.22), a trough in the TL graph will be obtained instead at $k_0 D_1/2 = 3.4865$.

In Fig. 4.15a, however, the 1-D result was obtained for the configuration with the end-offset port 1 offset on the major-axis at $x_{(2,1)e} = 0.4338 D_1/2$ (see Table 4.1) which is highly comparable with the tuning value suggested by Eq. (4.22). Consequently, the 1-D model does not exhibit a trough, rather an attenuation peak is observed which is nearly coincident with the peak at the $(2, 1)e$ mode predicted by the 3-D approach. This result, therefore, demonstrates that highly eccentric ellipses with aspect-ratio $D_2/D_1 = 0.35$ or smaller can indeed be double-tuned using the transverse plane wave model notwithstanding a very small difference between the exact and approximate values of the offset distances specified by the 3-D and 1-D models, respectively. Note that the second transverse trough frequency $k_0 D_1/2 = 3.4865$ is highly comparable with the $(2, 1)e$ mode resonance of an elliptical section with $D_2/D_1 = 0.35$ and smaller, refer to Tables 2.1–2.8. This further indicates the validity of the transverse plane wave model up to the $(2, 1)e$ mode resonance for highly eccentric ellipses. On the other hand, if the end port 1 is located at an offset distance given by Eq. (4.22), the 3-D approach will predict an attenuation peak in the immediate vicinity of the $(2, 1)e$ mode resonance; despite a slight mistuning, a broadband TL performance will still be observed that will be in a close agreement with the 1-D transverse plane wave prediction.

Figure 4.15b shows a noticeable mismatch between the attenuation peaks occurring near the $(2, 1)e$ mode predicted by the 3-D and 1-D approaches. Regardless of this deviation, the TL graph obtained by the transverse plane wave model can still be considered as a good approximation up to the $(2, 1)e$ mode for chambers with $D_2/D_1 = 0.5$, beyond this frequency it fails completely. The mismatch between the attenuation peaks near the $(2, 1)e$ mode can be eliminated by taking a slightly different offset location of the end port 1 in the 1-D model, i.e., one modifies the length of the side-branch cavity formed above the port 2. To this end, we iteratively find $y_{\text{effective1}}'$ such that root $\beta/2$ of the equation $F_1(y_{\text{effective1}}') = 0$, i.e., cavity resonance frequency is coincident with the $(2, 1)e$ mode resonance frequency. It was found that the effective non-dimensional cavity depth $y_{\text{effective1}}' = 0.28487$ which implies that the effective location of the end-offset port 1 from the center $x_{\text{effective1}} = 0.43026 D_1/2$. Note from Table 4.1 that the end port 1 is 'physically' offset at $x_{(2,1)e} = 0.4238 D_1/2$

on the major-axis, therefore, the value of effective location signifies that a shorter cavity depth is considered to account for a small right-shift in the attenuation peak. The end-correction length $\Delta_1 = (x_{(2,1)e} - x_{\text{effective1}}) = (0.43026 - 0.4238)D_1/2 = 0.00646D_1$ which indeed is only a small but crucial improvement. Similarly, for the end-centered port 2, the effective cavity depth $y'_{\text{effective2}} = 0.49684$ which signifies a smaller end-correction length $\Delta_2 = 0.00158D_1$, however, for this port, a larger effective cavity length is considered to account for the slight left-shift in the attenuation peak making it coincident with the peak at $(1, 1)e$ mode resonance. Note that considering simultaneous end-correction implies a small upward offset for ports 1 and 2 relative to their locations shown in Fig. 4.20b. In fact, this approach of introducing transverse end-corrections in the short end-chamber is indeed similar to the practice of tuning the extended-inlet and extended-outlet lengths of long chambers [24]. Figure 4.15b shows that the attenuation peaks in the TL graph obtained by the 1-D transverse plane wave model when the end-corrections were incorporated was coincident with the peaks predicted by the 3-D approach at the $(1, 1)e$ and $(2, 1)e$ mode resonance. This suggests that the end-corrected transverse plane wave approach can marginally improve the prediction up to the $(2, 1)e$ mode; however, it cannot deliver a double-tuned TL graph for chambers with $D_2/D_1 \approx 0.5$ due to large deviations from the 3-D approach.

Figure 4.16a shows a significant mismatch between the frequency of occurrence of the peak near the $(2, 1)e$ mode and a marginal deviation for the peak at $(1, 1)e$ mode resonance predicted by the 1-D and 3-D models. However, when end-corrections were incorporated, the attenuation peaks predicted by the 1-D model were coincident with the 3-D approach. Furthermore, the accuracy of the 1-D model was increased up to the $(2, 1)e$ mode resonance for straight-flow configuration; however, one cannot argue that this chamber is double-tuned. For the sake of documentation, it is mentioned that for such configuration with $D_2/D_1 = 0.65$, $y'_{\text{effective1}} = 0.28922$ and $y'_{\text{effective2}} = 0.49477$ which implies a greater upward offset relative to their physical locations.

For chamber with $D_2/D_1 = 0.85$, incorporation of end-corrections in the transverse plane model can deliver an improved performance only up to the $(1, 1)e$ mode resonance beyond which the 1-D results have no resemblance with the 3-D predictions, see Fig. 4.16b. Therefore, for ellipses approaching a circle, the validity of the transverse plane wave after incorporating appropriate transverse end-corrections can at most be extended up to the cut-on frequency of the first higher-order mode. Therefore, for a short circular chamber with an end-centered inlet (outlet) and end-offset outlet (inlet), it makes sense to introduce an end-correction offset only for the former. It was numerically found that the effective non-dimensional cavity depth is given by $y'_{\text{effective}} = 0.48892$, i.e., in the 1-D algorithm (or routines available with muffler designers), the centered port must be offset by a distance $0.01108D_1$ towards the same side as the offset port so that the peak predicted by the transverse plane wave is coincident with the peak at $(1, 0)$ mode resonance.

4.2.3 Inlet Port Centered and Outlet Located on the Minor-Axis

Often in engineering applications, the circular muffler shell is only slightly deformed to meet space constraints. For such low eccentricity elliptical sections, we explore the possibility of obtaining an alternate double-tuned configuration which can deliver a broadband attenuation comparable to that of a double-tuned configuration 2 analyzed in Sect. 4.2.1.1. A short elliptical chamber having an end-inlet port 1 centered on the intersection of the pressure nodal ellipse $\xi_{(0,2)e}$ of the $(0, 2)e$ mode and the minor-axis, and the end-outlet port 2 located at the ellipse center ($\delta_E = 0$) may possibly represent the desired muffler for small eccentricity values. The configuration 5 in Fig. 4.2c depicts the above arrangement of end ports. Note that the offset location of end-inlet port 1 along the minor-axis is given by Eq. (4.2) while for the end-outlet port 2, $\delta_{E2} = 0$.

Figure 4.21a and b presents the TL graphs for configuration 5 for aspect-ratio $D_2/D_1 = 0.95$ and 0.9, respectively. Here, results for both straight-flow and flow-reversal type chambers are included. The TL graphs shown in Fig. 4.21a are highly comparable with the results presented in Fig. 4.8a—a broadband attenuation is observed up to the resonance frequency of $(2, 2)e$ mode notwithstanding the inevitable occurrence of a 'trough' at the resonance frequency of $(2, 1)e$ or $(2, 1)o$ modes. Note that ≈ 10 dB attenuation is still produced at the trough which effectively extends the broadband attenuation range. The uplifted trough occurs because for nearly circular sections, the pressure nodal hyperbola $\eta_{(2,1)e}$ intersects the major-axis at nearly zero offset distance from the center, i.e., $x_{(2,1)e} \rightarrow 0$. (Refer to the $(2, 1)$ circumferential mode shape shown in Fig. 2.7b.) Figure. 4.21b shows that even for a small decrease in aspect-ratio, i.e., increase in eccentricity, we observe a substantial decrease in

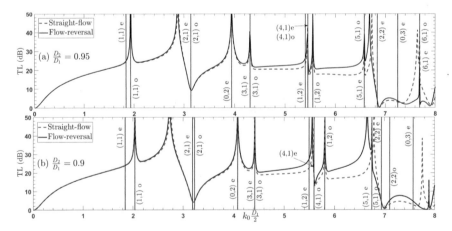

Fig. 4.21 TL performance of short flow-reversal and straight-flow elliptical end-chamber configurations having **a** $D_2/D_1 = 0.95$ and **b** $D_2/D_1 = 0.9$ with the end-inlet port centered and the end-outlet port located on the minor-axis at $\delta_{E2} = y_{(0,2)e}$

attenuation produced at the trough occurring near $(2, 1)e$ or $(2, 1)o$ mode resonances. Strictly speaking, for $D_2/D_1 = 0.9$, the broadband attenuation appears to break down at this frequency beyond which a good attenuation performance is observed up to the onset of $(2, 2)e$ mode. (Notice the difference in attenuation pattern between Figs. 4.21b and 4.7b.) For a still smaller aspect-ratio, e.g., $D_2/D_1 = 0.85$, it was easy to observe the collapse at the $(2, 1)e$ mode resonance beyond which the attenuation performance was rather inferior to its counterpart configuration 2 having the same aspect-ratio. In light of the above discussion, it is concluded that only for very small eccentricities, a short elliptical muffler having one port located at the center and other offset on the minor-axis can deliver a broadband attenuation performance comparable to its double-tuned counterpart.

4.2.4 Ports Offset at Equal Distances on the Major-Axis or Minor-Axis: Non-tuned Configuration

The configurations 3 and 4 shown in Fig. 4.2c depict elliptical mufflers with end ports 1 and 2 located at equal offset distances $\delta_{E1} = \delta_{E2} = \delta_0$, on the minor-axis and major-axis, respectively, at opposite ends, i.e., $(\theta_{E1}, \theta_{E2}) = (3\pi/2, \pi/2)$ for configuration 3 while $(\theta_{E1}, \theta_{E2}) = (0, \pi)$ for configuration 4. Furthermore, note that δ_0 corresponds to the pressure nodal ellipse of the $(0, 2)e$ mode and the pressure nodal hyperbola of the $(2, 1)e$ mode for configurations 3 and 4, respectively.

For both configurations 3 and 4, the TL graphs were computed for aspect-ratio $D_2/D_1 = 0.6$ though the results are not included here. The configuration 4 exhibited an expansion chamber type characteristics marked with troughs at the $(1, 1)e$ and $(3, 1)e$ modes because propagation of even-odd modes cannot be suppressed by either end ports, thereby implying an early breakdown or collapse of the TL spectrum, though a somewhat high attenuation was observed beyond the $(1, 1)e$ mode resonance up to frequencies slightly greater than the $(2, 1)e$ mode resonance. Nevertheless, in view of the relatively low attenuation, the overall TL performance of this configuration is rather poor, and such a non-tuned configuration is not effective from the acoustic design point-of-view. Furthermore, note that the 1-D transverse plane wave approach is able to accurately predict the expansion chamber type nature up to frequencies slightly greater than the $(1, 1)e$ mode [18, 19]. Similar comments apply for a short circular chamber with end ports offset at equal distances and located on opposite sides—here, the trough occurring at the $(1, 0)$ mode resonance leads to a breakdown of the attenuation pattern.

The configuration 3 also exhibits a simple expansion chamber type of behavior; however, a somewhat flat attenuation graph was observed for a greater frequency range which was followed by a trough at the $(1, 1)o$ mode resonance leading to the collapse of the TL spectrum. Note that for this configuration, the propagation of odd-odd and even-even modes cannot be suppressed as both ports are offset on the minor-axis, which implies that the 1-D transverse plane wave approach is not valid. The configuration 3 is also non-tuned and not recommended from the acoustic design perspective.

4.3 End-Inlet and Side-Outlet Configurations

In some configurations of single-/multi-pass perforated tube mufflers such as the one shown in Fig. 4.22, the short end-chamber has a port located on the side surface due to logistic reasons, while the other port is located on the end face at any arbitrary location. In this section, the TL performance of a short chamber with optimal locations of the end-inlet and side-outlet port is analyzed which is expected to deliver a broadband attenuation. Figure 4.22 schematically shows such four optimal configurations—For the muffler configuration 1, the end port is located at the center of the ellipse ($\delta_E = 0$), while for configuration 2, it is offset on the minor-axis. For both configurations 1 and 2, the side-outlet port is centered on the major-axis, i.e., $\eta_S = 0$. For the muffler configuration 3, the end port is located at $\delta_E = 0$, while for configuration 4, it is offset on the major-axis. In both configurations 3 and 4, the side-outlet port is centered on the minor-axis, i.e., $\eta_S = \pi/2$.

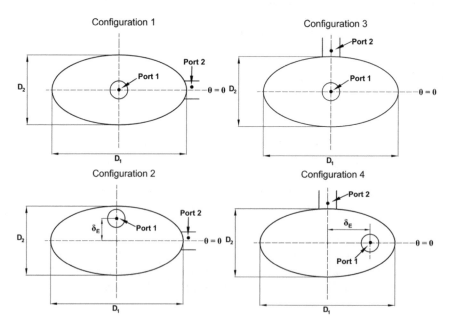

Fig. 4.22 End face of an elliptical muffler showing different angular locations of the end-inlet port 1 and the side-outlet port 2

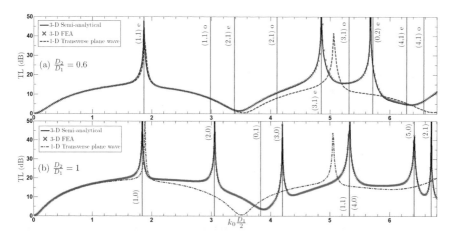

Fig. 4.23 TL performance of short end-centered inlet and side-outlet ($l_S = 0.5L$) configuration of **a** elliptical cylindrical geometry having $D_2/D_1 = 0.6$, $\eta_S = 0$ and **b** circular cylindrical geometry. The 3-D semi-analytical model is corroborated by comparing against the 3-D FEA results, while the 1-D transverse plane wave model is shown to breakdown at the resonance frequency of **a** $(2, 1)e$ mode and **b** $(1, 0)$ mode

Figure 4.23a and b presents the TL performance of end-inlet and side-outlet short muffler configuration 1 having (a) an elliptical cross-section with $D_2/D_1 = 0.6$ and (b) a circular cross-section (limiting case), respectively [15]. An excellent agreement is observed between the 3-D semi-analytical method and 3-D FEA simulation which corroborates the prediction obtained using the piston-driven model for this configuration. (In this section, the TL graphs are computed over a smaller frequency range up to 3000 Hz, or equivalently, up to the non-dimensional frequency $0.5k_0 D_1 = 6.86$.) Of significance in the TL spectrum shown in Fig. 4.23a are the occurrence of peaks at the $(1, 1)e$ and $(3, 1)e$ mode resonances, and a trough at $(2, 1)e$ mode resonance which limits the broadband attenuation range. Due to the end-centered inlet, only the even-even modes are excited, while the even-odd modes are suppressed. As a result, all impedance parameters are finite at the $(1, 1)e$ and $(3, 1)e$ mode resonances except the Z_{22} parameter which tends to infinity because the side-outlet located on the major-axis excites both even-even and even–odd modes. A trough near the resonance frequency of the $(2, 1)e$ mode occurs because the side-outlet port cannot be centered on the pressure nodal hyperbola of this mode, and therefore, the $[\mathbf{Z}]$ matrix parameters tend to infinity implying that TL $\rightarrow 0$.

For the short circular configuration 1 analyzed in Fig. 4.23b, it can similarly be shown that the peak at the resonances of the circumferential modes and the cross-modes is due to the end-centered inlet. The occurrence of a trough near the resonance of the first radial $(0, 1)$ mode, however, leads to the breakdown of broadband attenuation, and it occurs because the side-outlet is not centered at the pressure node of this mode. The 1-D transverse plane wave model is shown to be in a good agreement with the 3-D approaches up to the resonance frequency of $(2, 1)e$ mode in Fig. 4.23a and just before the resonance frequency of $(1, 0)$ circumferential mode in Fig. 4.23b. It fails beyond these frequencies due to the reasons noted in Sect. 4.2.2.1. Nevertheless, both approaches show that a short elliptical chamber with an end-centered inlet and a side-outlet located on the major-axis has attenuation characteristics similar to that of a side-branch resonator. Furthermore, a comparison with Fig. 4.4a which has an enhanced broadband TL range explains the reason for using the term *single-tuning* for end-inlet and side-outlet configuration 1. On the other hand, Fig. 4.23b exhibits a larger broadband TL range, and its performance is somewhat similar to the double-tuned short circular chambers analyzed in Fig. 4.8b, c up to the resonance frequency of the $(2,0)$ circumferential mode.

Figure 4.24 compares the TL performance of short elliptical muffler configurations 2–4 shown in Fig. 4.22b. An elliptical cross-section of the same aspect-ratio $D_2/D_1 = 0.6$ is also considered for these configurations. For the configuration 2, the end port is located on the intersection of the pressure nodal ellipse $\xi_{(0,2)e}$ and the minor-axis, and the offset distance $\delta_E = 39$ mm. The TL graph for the configuration 2 is observed to be nearly identical with the counterpart results shown in Fig. 4.23a in the low-frequency region, the only difference being the presence of an additional attenuation peak at the $(1, 1)o$ mode resonance. Therefore, offsetting the end-inlet port on the minor-axis does not further enhance the broadband attenuation range, which suggests that the configuration 2 represents a single-tuned chamber. A somewhat similar result was observed for the muffler configuration 3; the broadband nature was shown to collapse at the onset of the $(2, 1)e$ mode resonance, thereby signifying single-tuning for such chambers. A noticeable feature in the TL spectrum for the configuration 3 is the absence of an attenuation peak at the $(1, 1)e$ mode resonance because the even-odd modes are not excited. Rather, one observes attenuation peaks slightly before and at the resonance frequency of the $(1,1)o$ circumferential mode which improves the attenuation performance in the frequency range between $(1, 1)e$ and $(1, 1)o$ modes.

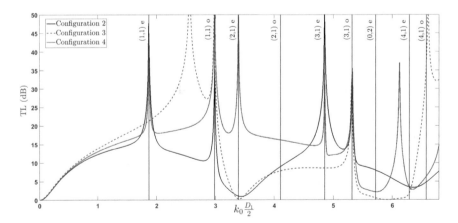

Fig. 4.24 TL performance of short elliptical muffler configurations 2–4 having an end-inlet and side-outlet illustrated in Fig. 4.22b

The configuration 4 exhibits a much improved attenuation performance resembling the double-tuned short chamber TL graphs presented in Fig. 4.4a. Note that in configuration 4, the end-inlet port is offset on the major-axis at pressure nodal hyperbola $\eta_{(2,1)e}$—the offset distance $\delta_E = 51.5$ mm. As a result, even-even, even–odd and odd-odd modes propagate which was the case for the configuration 2 presented in Fig. 4.2c. Therefore, quite expectedly, the TL graph presented in Fig. 4.24 for configuration 4 exhibits attenuation peaks at the $(1, 1)e$, $(1, 1)o$, $(3, 1)e$, $(3, 1)o$ circumferential modes, in addition to an attenuation peak at the $(2, 1)e$ mode resonance. The broadband attenuation nature, however, collapses at the onset of $(0, 2)e$ radial mode—this is unlike Fig. 4.4a in which a broadband attenuation up to the $(4, 1)e$ mode was obtained. However, for aspect-ratio $D_2/D_1 \approx 0.5$ or smaller, $(4, 1)e$ mode resonance occurs much earlier than the $(0, 2)e$ mode resonance. For such cases, the TL performance of configuration 4 is expected to be nearly the same as that of an end-inlet and end-outlet double-tuned configuration, e.g., Fig. 4.3a and b.

4.4 Design Guidelines

By virtue of the 3-D semi-analytical piston-driven approach, a parametric investigation was conducted to investigate the effect of location of inlet and outlet ports on the TL spectrum. The study has resulted in formulation of some design guidelines which recommend optimal port configurations for short elliptical and circular end-chamber mufflers that deliver a broadband TL performance for the maximum frequency range. Such double-tuned flow-reversal and straight-through muffler configurations are mentioned below.

1. Elliptical chambers with an end-inlet port offset on the major-axis at the pressure nodal hyperbola of $(2, 1)e$ mode and an end-outlet port offset on the minor-axis at the pressure nodal ellipse of $(0, 2)e$ mode.
2. Circular chambers with an end-inlet port located at the center of the circular section and an end-outlet port located at an offset distance corresponding to the pressure nodal circle of the $(0, 1)$ radial mode.

   From points 1 and 2 as well as from Fig. 4.10b, note that the recommend practice of designing a short elliptical muffler of high eccentricity is in clear contrast with the principle of designing a short circular chamber [14].
3. Elliptical chambers with aspect-ratio $D_1/D_1 \approx 0.5$ or smaller having an end-inlet port located at the ellipse center and an end-outlet port offset on the major-axis at the pressure nodal hyperbola of $(2, 1)e$ mode. The transverse plane wave model may also be used to double-tune the TL performance of this configuration with $D_2/D_1 = 0.35$ up to ≈ 0.4. For chambers having aspect-ratio $D_2/D_1 \approx 0.65$ or larger, the broadband pattern collapses at the onset of $(0, 2)e$ mode, but the performance tends to deteriorate beyond the $(2, 1)e$ mode resonance which occurs much earlier, thereby suggesting a single-tuned nature.

4. Low eccentricity elliptical chambers with aspect-ratio $D_2/D_1 \geq 0.95$ having an end-inlet port located at the ellipse center and an end-outlet port offset on the minor-axis at the pressure nodal ellipse of the $(0, 2)e$ mode. For $D_2/D_1 < 0.95$, the low trough occurring at the $(2, 1)e$ mode leads to a breakdown of the broadband attenuation resulting in a single-tuned chamber.

For end-inlet and side-outlet elliptical mufflers, the following configuration can be double-tuned:
5. Chambers with aspect-ratio $D_2/D_1 \approx 0.6$ or smaller having an end-inlet offset on the major-axis at the pressure nodal hyperbola of $(2, 1)e$ mode and a side-outlet port located on the minor-axis, i.e., at $\eta_S = \pi/2$.

Note that elliptical muffler configurations with the side-outlet located on the major-axis and the end-inlet either centered or offset on the minor-axis are single-tuned. Similar comment holds for (a) elliptical mufflers with the side-outlet located on minor-axis and end-inlet located at ellipse center as well as for (b) circular muffler configuration when the included angle between the end-offset inlet and end-offset outlet or side-outlet is 90°.

References

1. D.F. Ross, A finite element analysis of perforated component acoustic systems. J. Sound Vib. **79**, 133–143 (1981)
2. N.S. Dickey, A. Selamet, J.M. Novak, Multi-pass perforated tube silencers: A computational approach. J. Sound Vib. **211**, 435–448 (1998)
3. M.L. Munjal, Analysis of a flush-tube three-pass perforated element muffler by means of transfer matrices. Int. J. Acoust. Vib. **2**, 63–68 (1999)
4. M.L. Munjal, Analysis of extended-tube three-pass perforated element muffler by means of transfer matrices, in *Proceedings of the Fifth International Congress on Sound and Vibration (ICSV 1997)*, Adelaide, SA, Australia
5. A. Selamet, V. Easwaran, A.G. Falkowki, Three-pass mufflers with uniform perforations. J Acoust. Soc. Am. **105**, 1548–1562 (1999)
6. Z.L. Ji, A. Selamet, Boundary element analysis of three-pass perforated duct mufflers. Noise Control Eng. J. **48**, 151–156 (2000)
7. Z.L. Ji, F.Z. Fang, Three-pass perforated tube muffler with end-resonator. SAE technical 2011-01-1529 (2011)
8. Y. Fan, Z. Ji, Three-pass perforated tube muffler with perforated bulkheads. Adv. Mech. Eng. **8**, 1–11 (2016)
9. H. Huang, Z. Ji, Z. Li, Influence of perforation and sound-absorbing material filling on acoustic attenuation performance of three-pass perforated mufflers. Adv. Mech. Eng. **10**, 1–11 (2018)
10. A. Verma, M.L. Munjal, Flow-acoustic analysis of the perforated-baffle three-chamber hybrid muffler configurations. SAE Int. J. Passeng. Cars—Mech. Syst. **8**, 370–381 (2015)
11. A. Mimani, M.L. Munjal, Acoustic end-correction in a flow-reversal end chamber muffler: a semi-analytical approach. J. Comput. Acoust. **24**, 1650004 (2016)
12. Z. Fang, Z.L. Ji, Acoustic attenuation analysis of expansion chambers with extended inlet/outlet. Noise Control Eng. J. **61**, 240–249 (2013)
13. C.D. Gaonkar, D.R. Rao, K.M. Kumar, M.L. Munjal, End corrections for double-tuning of the same-end inlet-outlet muffler. Appl. Acoust. **159**, 107116 (2020)

14. A. Selamet, Z.L. Ji, Acoustic attenuation performance of circular flow-reversing chambers. J. Acoust. Soc. Am. **104**, 2867–2877 (1998)
15. A. Mimani, M.L. Munjal, 3-D acoustic analysis of elliptical chamber mufflers having an end inlet and a side outlet: an impedance matrix approach. Wave Motion **49**, 271–295 (2012)
16. A. Selamet, F.D. Denia, Acoustic behavior of short elliptical chambers with end central inlet and end offset or side outlet. J. Sound Vib. **245**, 953–959 (2001)
17. P.C.-C. Lai, W. Soedel, Two-dimensional analysis of thin, shell or plate like muffler elements. J. Sound Vib. **194**, 137–171 (1996)
18. A. Mimani, M.L. Munjal, Transverse plane wave analysis of short elliptical chamber mufflers: an analytical approach. J. Sound Vib. **330**, 1472 (2011)
19. A. Mimani, M.L. Munjal, Transverse plane-wave analysis of short elliptical end-chamber and expansion-chamber mufflers. Int. J. Acoust. Vib. **15**, 24 (2010)
20. A. Selamet, Z.L. Ji, Diametral plane-wave analysis for short circular chambers with end offset inlet/outlet and side extended inlet/outlet. J. Sound Vib. **214**, 580 (1998)
21. R.D. Cook, D.S. Malkus, M.E. Plesha, R.J. Witt, *Concepts and Applications of the Finite Element Analysis* (Wiley, New York, 2001)
22. A. Mimani, M.L. Munjal, Acoustical behavior of single inlet and multiple outlet elliptical cylindrical chamber muffler. Noise Control Eng. J. **60**, 605–626 (2012)
23. F.D. Denia, J. Albelda, F.J. Fuenmayor, A.J. Torregrosa, Acoustic behaviour of elliptical chamber mufflers. J. Sound Vib. **241**, 401 (2001)
24. P.C. Chaitanya, M.L. Munjal, Tuning of the extended concentric tube resonators. Int. J. Acoust. Vib. **16**, 111–118 (2011)
25. A. Selamet, Z.L. Ji, Acoustic attenuation performance of circular expansion chambers with offset inlet/outlet: I. Analytical approach. J. Sound Vib. **213**, 601–617 (1998)
26. M.V. Lowson, S. Baskaran, Propagation of sound in elliptic ducts. J. Sound Vib. **38**, 185–194 (1975)
27. M.L. Munjal, *Acoustics of Ducts and Mufflers*, 2nd edn. (Wiley, Chichester, UK, 2014).
28. D.T. Blackstock, *Fundamentals of Physical Acoustics*, Chap. 3 (Wiley, New York, 2000)
29. https://www.maplesoft.com/products/maple/
30. L.E. Kinsler, A.R. Frey, A.B. Coppens, J.V. Sanders, *Fundamentals of Acoustics* (Wiley, New York, 2000).
31. V.H. Gupta, V. Easwaran, M.L. Munjal, A modified segmentation approach for analyzing plane wave propagation in non-uniform ducts with mean flow. J. Sound Vib. **182**, 697–707 (1995)

Chapter 5
Summary of Contribution and Directions for Future Work

5.1 Contribution

This monograph has tabulated for the first time, the parametric zeros and the non-dimensional resonance frequencies of the higher-order transverse modes of a rigid-wall elliptical waveguide for a complete range of aspect-ratio. This is followed by an analysis of the first few mode shapes. Based on the modal summation and uniform piston-driven model, a theoretical formulation is presented for characterizing and evaluating the TL performance of reactive elliptical muffler configurations having an end-inlet and end-/side-outlet [1, 2]. The TL analysis has resulted in formulation of a set of comprehensive guidelines for designing short elliptical and circular mufflers which are used as end-chambers in a modern-day automotive exhaust system. The guidelines suggest optimal port locations which suppress the propagation of certain higher-order transverse modes, thereby yielding a flat but broadband attenuation despite a small expansion volume. More precisely, it was shown that locating the inlet and outlet ports on the major-axis and minor-axis, respectively, at appropriate pressure nodes yields a broadband attenuation for Helmholtz number $0.5k_0 D_1 > 6$. Similarly, for short chambers of nearly circular cross section, a concentric end-inlet and offset end-outlet centered on the pressure node of the first radial mode yield a broadband attenuation for Helmholtz number as high as $0.5k_0 D_1 \approx 7$, see Ref. [3].

The convective and dissipative effect of mean flow, however, have been ignored as a stationary medium was considered for analysis. Nonetheless, for typically low Mach number flow $M_0 \leq 0.15$ as is the case in exhaust/tail pipes of automotive silencers, the basic nature of TL graph is expected to remain unaltered for reactive silencer configurations considered here. Therefore, the design guidelines enumerated at the end of Chap. 4 based on the assumption of a stationary medium will also be valid for low Mach number flows. This is because the transverse mode shapes remain same even in the presence of flow, except that the associated axial wave numbers incorporate the convective effect. However, for perforated reactive silencers, mean flow is known to somewhat reduce the low-frequency attenuation and dramatically bring down the attenuation peaks produced at axial resonances [4–6], while for

A. Mimani, *Acoustic Analysis and Design of Short Elliptical End-Chamber Mufflers*, https://doi.org/10.1007/978-981-10-4828-9_5

dissipative silencers, effect of mean flow is to lower the TL graph for most frequencies [7, 8]. Furthermore, the troughs are slightly lifted due to flow, see the discussion in Munjal [9], while the attenuation increases above the cut-on frequency [4].

5.2 Future Investigation: Analysis and Design of Perforated Elliptical Mufflers for Industrial and Automotive Applications

Often in an automotive muffler design, it is a common practice to use a straight-through perforated pipe (airway) configuration which allows the acoustic waves to interact with the annular cavity through the perforate holes [4–8, 10–13]. The benefit of using a perforated airway is that it guides the mean flow, thereby preventing its direct expansion in the chamber (at inlet), thus eliminating the possibility of flow-induced or aeroacoustic noise at sudden area changes due to the formation of free-shear layers. Consequently, the airway significantly reduces the backpressure.

Figures 4.1 and 5.1 show some perforated duct muffler configurations having an elliptical cross section. Figures 4.1a, c and 5.1f depict multi-pass flow-reversal configurations, while Fig. 4.1b, d show single-pass straight-flow configurations. Figure 5.1a–d shows straight-through grazing-flow muffler configurations having a concentric perforated airway, and Fig. 5.1e shows a cross-flow configuration having one concentric and other eccentric open-ended perforated airway.

While the 3-D semi-analytical formulation is suited for evaluating the TL performance of empty rigid-wall elliptical chamber muffler configurations without port extensions [1, 2, 21, 22], the configurations shown in Figs. 4.1 and 5.1 cannot be analyzed in terms of Mathieu function modal expansions as the transverse modes (eigenfunctions) cannot be determined analytically. Therefore, numerical techniques which can model non-dissipative/dissipative perforated mufflers in a computationally efficient manner must be used; this will be the focus of future investigations involving elliptical cylindrical chambers.

One such popular technique is the numerical mode-matching (NMM) method which uses the FEA to compute the 2-D transverse modes (eigenfunctions) and the associated resonance frequencies (eigenvalues) within the arbitrary but uniform cross-section chamber as well as in the circular airway [7, 8, 10–14]. By enforcing the continuity condition of acoustic velocity and pressure fields at the port-chamber interface, the NMM generates a system of linear algebraic equations which is to be solved for the unknown modal amplitudes in terms of the incident wave amplitude (inlet), leading to the computation of TL. In essence, the NMM reduces a full 3-D FEA problem to its 2-D counterpart and is especially suited for efficiently evaluating TL performance of large industrial silencers such as the one shown in the photograph in Fig. 5.2, see also Mimani and Kirby [23]. These elliptical (circular) silencers typically use a straight-through airway pipe as shown in Fig. 5.1a–d, and such silencer configurations offer an advantage of low-backpressure. The major-axis, chamber

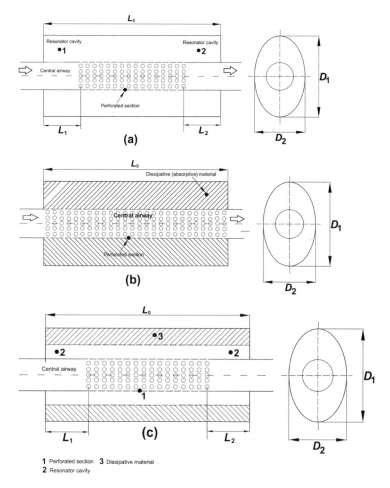

1 Perforated section   3 Dissipative material
2 Resonator cavity

Fig. 5.1 a Empty elliptical silencer having a partially perforated concentric duct (airway) which forms quarter-wave resonators due to extended-inlet and extended-outlet, see Refs. [4–6]. An elliptical chamber muffler with extended-inlet and extended-outlet but without a perforated bridge is analyzed in Ref. [14], while its circular counterpart with concentric length extensions was studied in Ref. [15], **b** Dissipative elliptical silencer having a perforated concentric airway. The annular cavity is filled with sound-absorbing material, see Refs. [7, 8, 10, 11, 13, 16], **c** Elliptical silencer having a partially perforated concentric airway (1) with an annular air-gap (empty resonator cavity (2)) between the airway and dissipative material (3), see Ref. [17]. The dissipative material is covered with a partially perforated screen. Such configurations are also referred to as hybrid mufflers [9], **d** Triple-chamber elliptical silencer having a concentric perforated airway with mean flow. The resonator chamber C1 is empty, while resonator chambers C2 and C3 are filled with dissipative material, see Refs. [7, 18]. In this configuration, the chambers C1, C2 and C3 are typically used to attenuate low, mid and high-frequency noise, respectively, **e** Three-duct, cross-flow elliptical silencer having overlapping open-end perforated tubes, analyzed in Ref. [19], **f** Three-pass elliptical silencer having a complex internal structure: system of perforated tubes, cross-baffles with perforates (bulkhead) and long end-chambers with tube extensions (forming resonator cavities). A variant of this configuration is analyzed in Ref. [20]

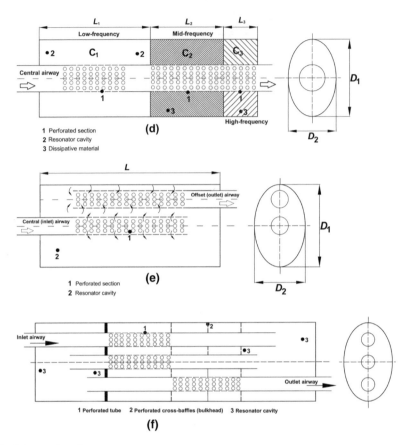

Fig. 5.1 (continued)

length and airway diameter of the industrial silencers can be as large as 1000 mm, 2000 mm and 350 mm, respectively. For such large cross-dimensions, it is necessary to consider modal coupling in the chamber and port (airway) to obtain accurate predictions even at low frequencies. The design challenge here is to deliver high levels of noise attenuation at low frequencies below 150 Hz in a relatively confined space. This is because large internal combustion engines (up to 60 L capacity) used in mining/excavation trucks and dozers, and freight trains produce high levels of noise in very low frequency range. Therefore, one must rely on reactive elements such as empty chambers with extensions at inlet and outlet (Fig. 5.1a, c) to attenuate low-frequency noise as dissipative elements are ineffective at such frequencies. Of course, it is assumed that at higher frequencies, dissipative mufflers such as Fig. 5.1b, d will be used to extend the noise attenuation to the full frequency range of interest. Often in such applications, limited space is available in one direction only which is a common design constraint, and use of elliptical mufflers is an attractive option

Fig. 5.2 Photograph of a
large industrial silencer of
elliptical cylindrical
geometry, typically used for
muffling noise during mining
and ground excavation
operations. (The photograph
was taken from the website
https://www.minetek.com/
en/industrial-exhausts/ with
due permission obtained
from *Minetek*, New South
Wales, Australia)

because it has the potential to deliver higher area ratios (for targeting low frequencies)
under such physical constraints.

In Mimani and Kirby [23], some preliminary design guidelines based on the NMM
technique were presented for delivering a broadband attenuation for large straight-
through single-chamber reactive silencer configuration shown in Fig. 5.1a. The extension lengths at the inlet and outlet were taken to be slightly smaller than half-chamber
and quarter-chamber lengths, respectively, so that the troughs in the attenuation graph
at the first and second chamber resonance frequencies were minimized. In particular,
it was shown that regardless of the elliptical or circular cross-section, a concentric
airway is preferred for delivering a wider range of broadband performance. This
is because for a circular silencer, a concentric airway will excite only the axisymmetric modes, and for an elliptical silencer, only those modes are excited which are
symmetrical about both major-axis and minor-axis. Further, offsetting the airway on
the minor-axis may be desirable to preserve the effective attenuation range up to the
mid-frequency. For automotive applications, the muffler size is much smaller, and
typically, only planar waves propagate in the airway throughout the frequency range.
In such situation, one can fine-tune (or optimize) the extension lengths through a
trial-and-error method so that resonance peak due to quarter-wave resonators at inlet
and outlet exactly nullifies the troughs at the first three chamber resonance frequencies delivering a wide-band attenuation. Developing expressions for end-correction
to tune the inlet/outlet extension lengths based on NMM (and experimental vali-
dation thereof) for muffler configuration shown in Fig. 5.1a as a general function
of the aspect-ratio, airway diameter and offset distance and porosity is therefore an
important future work.

In the aforementioned work, a stationary medium was assumed implying that dissipative effect of mean flow at perforate-chamber junction as well as its convective effect was not taken into consideration. Although the design guidelines will remain unaltered in presence of flow, to obtain more realistic estimates of the attenuation spectrum for industrial silencers, it is necessary that future investigations should take flow into consideration. The convective effect can be accounted for by considering additional terms in the eigenvalue problem and using appropriate matching conditions at the port-chamber interface [10], whereas the more significant dissipative effect can be modeled through the (complex) perforate impedance wherein the nonzero grazing flow substantially increases the resistive part, thereby inducing damping into the muffler system [20, 24]. Additionally, the large silencers may be subject to non-planar excitation due to incident disturbances at the inlet from exhaust pipe of internal combustion engines, as is often the case with turbo-machinery application, see Tyler and Sorfin [25] or Mechel [26]. An analysis of the attenuation performance under such conditions, especially above the first cut-on frequency using NMM and the equal modal energy density (EMED) approach [26], presents an interesting avenue for further work on large muffler configurations shown in Fig. 5.1a–d, see Ref. [27].

The problem of analyzing multi-pass perforated tube muffler configurations for automotive applications using the NMM technique presents an important design challenge. Some of the representative configurations are shown in Figs. 4.1 and 5.1e, f. Here, a high acoustic attenuation performance needs to be delivered in a constrained space and a significantly high backpressure can be tolerated. Under such requirements, one is forced to use multiple perforated tubes in a cross-flow or bias-flow condition (Fig. 5.1e, f) which significantly increases the resistive part of perforate impedance, thereby enhancing the attenuation due to dissipative loss. The modeling challenge here is to first employ FEA to obtain the transverse modes of the muffler cross section having multiple tubes and annular region interacting through perforate boundaries. Next, the modes corresponding to two adjacent cross sections need to be matched, and due to the complex internal structure of multi-pass mufflers, there are several matching planes resulting in a large system of linear simultaneous equations. While the associated book-keeping is certainly tedious, the main issue here is the stability of the solution set; the coefficient matrix tends to become close to singular when flow is present, thereby presenting difficulty in the straightforward application of NMM. The development of computational techniques to circumvent such potential problems will be a key focus, and this is important because NMM allows one to efficiently conduct parametric studies required for designing such systems. One such possible alternative is to combine the numerical (FEA) and analytical modal analysis during mode-matching for analyzing the configurations shown in Fig. 4.1; while the modes of the perforate-chamber sub-system can be obtained using FEA, the transverse modes of short end-chamber can be obtained analytically followed by matching across their interfaces. Such a technique can help us readily analyze the effect of optimal end port location (Chap. 4) on the overall attenuation performance of such muffler systems. The short end-chambers which have a pass-tube such as the one shown in Fig. 1.1b, however, cannot be analyzed using an analytical modal summation approach; for such multi-pass muffler configurations, the NMM method has to be used for all sub-systems. Furthermore, it is required to investigate the effect of tuning the extension lengths of pipes in the two end-chambers of the configuration shown in Fig. 5.1f on its overall TL performance.

References

1. A. Mimani, M.L. Munjal, 3-D acoustic analysis of elliptical chamber mufflers having an end inlet and a side outlet: an impedance matrix approach. Wave Motion **49**, 271–295 (2012)
2. A. Mimani, M.L. Munjal, Acoustical behavior of single inlet and multiple outlet elliptical cylindrical chamber muffler. Noise Control Eng. J. **60**, 605–626 (2012)
3. A. Selamet, Z.L. Ji, Acoustic attenuation performance of circular flow-reversing chambers. J. Acoust. Soc. Am. **104**, 2867–2877 (1998)
4. Z.L. Ji, H.S. Xu, Z.X. Kang, Influence of mean flow on acoustic attenuation performance of straight-through perforated tube reactive silencers and resonators. Noise Control Eng. J. **58**, 12–17 (2010)
5. E. Ramya, M.L. Munjal, Improved tuning of the extended concentric tube resonator for wide-band transmissions loss. Noise Control Eng. J. **62**, 252–263 (2014)
6. P.C. Chaitanya, M.L. Munjal, Tuning of the extended concentric tube resonators. Int. J. Acoust. Vibr. **16**, 111–118 (2011)
7. Z. Fang, Z.L. Ji, Numerical mode matching approach for acoustic attenuation predictions of double-chamber perforated tube dissipative silencers with mean flow. J. Comput. Acoust. **22**, 1450004 (2014)
8. R. Kirby, Transmission loss predictions for dissipative silencers of arbitrary cross section in the presence of mean flow. J. Acoust. Soc. Am. **114**, 200–209 (2003)
9. M.L. Munjal, *Acoustics of Ducts and Mufflers*, 2nd edn. (Wiley, Chichester, UK, 2014)
10. R. Kirby, A comparison between analytic and numerical methods for modelling automotive dissipative silencers with mean flow. J. Sound Vibr. **325**, 565–582 (2009)
11. J. Albelda, F.D. Denia, M.I. Torres, F.J. Fuenmayor, A transversal substructuring mode matching method applied to the acoustic analysis of dissipative mufflers. J. Sound Vibr. **303**, 614–631 (2007)
12. F.D. Denia, A. Selamet, F.J. Fuenmayor, R. Kirby, Acoustic attenuation performance of perforated dissipative mufflers with empty inlet/outlet extensions. J. Sound Vibr. **302**, 1000–1017 (2007)
13. A.G. Antebas, F.D. Denia, A.M. Pedrosa, F.J. Fuenmayor, A finite element approach for the acoustic modeling of perforated dissipative mufflers with non-homogeneous properties. Math. Comput. Model. **57**, 1970–1978 (2013)
14. Z. Fang, Z.L. Ji, Acoustic attenuation analysis of expansion chambers with extended inlet/outlet. Noise Control Eng. J. **61**, 240–249 (2013)
15. A. Selamet, Z.L. Ji, Acoustic attenuation performance of circular expansion chambers with extended inlet/outlet. J. Sound Vibr. **223**, 197–212 (1999)
16. A. Selamet, I.J. Lee, N.T. Huff, Acoustic attenuation of hybrid silencers. J. Sound Vibr. **262**, 509–527 (2003)
17. N.K. Vijayasree, M.L. Munjal, On an Integrated Transfer Matrix method for multiply connected mufflers. J. Sound Vibr. **331**, 1926–1938 (2012)
18. A. Selamet, F.D. Denia, A.J. Besa, Acoustic behavior of circular dual-chamber mufflers. J. Sound Vibr. **265**, 967–985 (2003)
19. G.R. Gogate, M.L. Munjal, Analytical and experimental aeroacoustic studies of open-ended three-duct perforated elements used in mufflers. J. Acoust. Soc. Am. **97**, 2919–2927 (1995)
20. T. Elnady, M. Abom, S. Allan, Modeling perforates in mufflers using two-ports. J. Vibr. Acoust. **15**, 24–38 (2010)

21. F.D. Denia, J. Albelda, F.J. Fuenmayor, A.J. Torregrosa, Acoustic behaviour of elliptical chamber mufflers. J. Sound Vibr. **241**, 401–421 (2001)
22. A. Mimani, M.L. Munjal, Acoustic end-correction in a flow-reversal end chamber muffler: a semi-analytical approach. J. Comput. Acoust. **24**, 1650004 (2016)
23. A. Mimani, R. Kirby, Design of large reactive silencers for industrial applications, in Proceeding. of InterNoise, 26–29 August, Chicago, USA (2018)
24. C. Lahiri, F. Bake, A review of bias flow liners for acoustic damping in gas turbine combustors. J. Sound Vibr. **400**, 564–605 (2017)
25. J.M. Tyler, T.G. Sofrin, Axial flow compressor noise studies. SAE Trans. **70**, 309–32 (1962)
26. F.P. Mechel, Theory of baffle-type silencers. Acustica **70**, 93–111 (1990)
27. P. Williams, R. Kirby, J. Hill, M. Abom, C. Malecki, Reducing low frequency tonal noise in large ducts using a hybrid reactive-dissipative silencer. Appl. Acoust. **131**, 61–69 (2018)

Index

© The Author(s), under exclusive license to Springer Nature Singapore Pte Ltd. 2021
A. Mimani, *Acoustic Analysis and Design of Short Elliptical End-Chamber Mufflers*,
https://doi.org/10.1007/978-981-10-4828-9

Printed in the United States
By Bookmasters